Ali Khonya

System-Level Modeling of the Powertrain for an Electric Aircraft with Superconducting Components

Karlsruher Schriftenreihe zur Supraleitung Band 039

HERAUSGEBER

Prof. Dr. Tabea Arndt
Prof. Dr. rer. nat. Bernhard Holzapfel
Prof. Dr. rer. nat. Sebastian Kempf
Prof. Dr.-Ing. Mathias Noe

Eine Übersicht aller bisher in dieser Schriftenreihe erschienenen Bände finden Sie am Ende des Buches.

System-Level Modeling of the Powertrain for an Electric Aircraft with Superconducting Components

by
Ali Khonya

KIT Scientific Publishing

Karlsruher Institut für Technologie
Institut für Technische Physik

System-Level Modeling of the Powertrain for an
Electric Aircraft with Superconducting Components

Zur Erlangung des akademischen Grades eines Doktors der Ingenieurwissenschaften (Dr.-Ing.) von der KIT-Fakultät für Elektrotechnik und Informationstechnik des Karlsruher Instituts für Technologie (KIT) genehmigte Dissertation

von Ali Khonya, M.Sc.

Tag der mündlichen Prüfung: 17. März 2025
Hauptreferent: Prof. Dr.-Ing. Mathias Noe
Korreferent: Prof. Dr.-Ing. Antonio Morandi

Impressum

KIT Scientific Publishing

Karlsruher Institut für Technologie (KIT)
KIT Scientific Publishing
Straße am Forum 2
D-76131 Karlsruhe

KIT Scientific Publishing is a registered trademark
of Karlsruhe Institute of Technology.
Reprint using the book cover is not allowed.

www.bibliothek.kit.edu/ksp.php | E-Mail: info@ksp.kit.edu | Shop: www.ksp.kit.edu

Print on Demand 2025 – Gedruckt auf FSC-zertifiziertem Papier

ISSN 1869-1765
ISBN 978-3-7315-1435-0
DOI 10.5445/KSP/1000181030

Acknowledgments

Firstly, I would like to express my heartfelt appreciation to my supervisor, Prof. Dr.-Ing. Mathias Noe, the executive director of the Institute for Technical Physics (ITEP), where I did my Ph.D., for always supporting and encouraging me and providing excellent guidance throughout my Ph.D. work. His expertise and mentorship were significant assistance in shaping this work by offering productive and helpful insights that enhanced the quality of my work. His openness to curiosity and learning inspired my passion for this work. Beyond his professional guidance, I sincerely appreciate him for providing valuable feedback that benefited me in my career and personal growth. His continuous support since the very first day made this journey enjoyable and fulfilling.

I would like to thank Prof. Dr.-Ing. Giovanni De Carne, the head of the Real-time Simulation group and Energy Lab director, of which I was a part. The friendly atmosphere and the flexible working environment he created were solid reasons for me to enjoy my work and show my capabilities. I also thank the colleagues at Airbus UpNext, especially Dr. Frederick Berg, for supporting me in this project. Many thanks as well to Prof. Antonio Morandi for his insights to improve this work.

I am also grateful for all the support I got from my colleagues at ITEP and the RTSET group: Wescley, Asef, Fargah, Masoome, Maëva, Gabriele, Mahshid, Nima, Karthik, Qiucen, Felix, Sarah, Michael, Diego, Honeymol , Christian, Ruslan, Stelios, and Pongbo. I especially thank Nils, my perfect officemate and a great friend, someone who was always there in the office to support and encourage me. My colleagues are a big part of my Ph.D., making it the best career experience of my life. Thank you all!

I am deeply thankful to my family, especially my parents; they sacrificed a lot to help me reach where I am. The love and support they provided for me are invaluable. Many thanks as well to my dear friends, Masoud and Shadi, who constantly supported me. Lastly, I express deep appreciation to my beloved Golnaz. This work without her would be impossible; therefore, I dedicate it entirely to her for consistently supporting and encouraging me.

Kurzfassung

Der Luftverkehr ist für etwa 2,4 % der jährlichen weltweiten Kohlenstoffemissionen verantwortlich. Ohne weitere Maßnahmen werden sich die CO_2-Emissionen bei einem jährlichen Wachstum des Passagieraufkommens von 3,6-3,8 % bis 2050 voraussichtlich verdoppeln. Diese beträchtlichen Auswirkungen wurden in den letzten Jahrzehnten in Angriff genommen, und es wurden mehrere Projekte wie der EU-Flightpath 2050 ins Leben gerufen, um die Herausforderung eines kohlenstofffreien Flugverkehrs zu bewältigen. Um dieses Ziel zu erreichen, wurden verschiedene Alternativen zu herkömmlichem Flugbenzin untersucht. Die praktikabelsten Lösungen für den Energieträger der Zukunft sind nachhaltige Flugkraftstoffe (SAF) und Wasserstoff. Im Gegensatz zu SAF, das nur in Gasturbinen verbrannt werden kann, kann grüner Wasserstoff sowohl in Verbrennungsmotoren als auch zur Stromerzeugung durch Brennstoffzellenstacks in einem Elektroflugzeug verwendet werden. Flüssiger Wasserstoff, LH_2, hat einen fast dreimal höheren Energiegehalt pro Masseneinheit als Flugzeugtreibstoff, was ihn zu einer idealen Wahl für die Dekarbonisierung der Luftfahrt macht. Mehrere Studien haben LH_2 als die langfristige Lösung für Langstreckenflüge identifiziert, die SAF ersetzen soll. Die Verwendung von Wasserstoff in einem Elektroflugzeug auf Brennstoffzellenbasis hat zahlreiche Vorteile gegenüber der Verbrennung in Gasturbinen, darunter ein höherer Wirkungsgrad, ein leiserer Betrieb und die Vermeidung von Emissionen wie NO_x und SO_2 sowie CO_2. Daher haben mehrere Unternehmen wie Airbus und Boeing bereits die Elektrifizierung der Luftfahrt in verschiedenen Konfigurationen erwogen, darunter vollelektrische, hybridelektrische und turboelektrische Flugzeuge. Aufgrund der vielen Vorteile konzentriert sich diese Arbeit auf ein vollelektrisches Flugzeug auf Brennstoffzellenbasis.

Mit dem Anstieg der erforderlichen Leistung nehmen Größe und Gewicht der Komponenten erheblich zu, insbesondere die Größe des Wasserstofftanks. Daher müssen Größe und Gewicht der Komponenten des Antriebsstrangs leichter und effizienter sein, da eine höhere Effizienz zu einem geringeren Wasserstoffverbrauch und folglich zu kleineren Tanks führt. Supraleitende Materialien stellen eine vielversprechende Lösung dar, da sie eine effizientere Stromübertragung bei geringerem Gewicht und Platzbedarf ermöglichen. Dies hat dazu geführt, dass diese Technologie bei mehreren Projekten zur Elektrifizierung der Luftfahrt, wie dem ASCEND-Projekt von Airbus, in Betracht gezogen wurde. Ein Elektroflugzeug mit einem supraleitenden Antriebsstrang bietet somit eine einzigartige Gelegenheit für einen schnellen Übergang zu einer kohlenstofffreien Luftfahrt. Ein solcher Antriebsstrang umfasst mehrere Komponenten, wie z. B. einen Brennstoffzellenstapel, einen supraleitenden Fehlerstrombegrenzer, DC/AC-Kabel und -Motor sowie einen DC/AC-Wechselrichter. Die genaue Modellierung des gesamten Antriebsstrangs und jeder einzelnen Komponente ist eine wichtige Voraussetzung für das Verständnis des stationären und transienten Betriebsverhaltens. In dieser Arbeit wird der Ansatz zur Modellierung eines solchen

Antriebsstrangs in MATLAB/SIMULINK für die Systemsimulation in verschiedenen Flugszenarien vorgestellt.

Ein Brennstoffzellenstapel als Energiequelle wandelt die chemische Energie des Wasserstoffs in elektrische Energie um. Da Protonenaustauschmembran-Brennstoffzellen (PEMFC) geeignete Eigenschaften für Luftfahrtanwendungen bieten, wird dieser Brennstoffzellentyp in dieser Arbeit betrachtet. Die Brennstoffzelle liefert die Leistung mit einer nichtlinearen Spannungs-Strom-Variation, d.h. je höher der Strom ist, desto niedriger ist die Ausgangsspannung. Diese Arbeit zeigt den Ansatz für die Entwicklung eines Brennstoffzellenstapels auf der Grundlage der Anforderungen und der Modellierung seines stationären und instationären Verhaltens. Die meisten Forschungsarbeiten gehen von einem resistiv-kapazitiven Einschwingverhalten der Brennstoffzelle aus. Die Experimente zeigen jedoch oft ein resistiv-induktives Verhalten oder, je nach Brennstoffzelle, eine Kombination aus beidem. In dieser Arbeit wird ein Ansatz zur Modellierung eines beliebigen Verhaltens vorgestellt.

Zur Vermeidung von Kurzschlussströmen im Antriebsstrang wird ein resistiver supraleitender Fehlerstrombegrenzer (RSFCL) integriert, der Hochtemperatur-Supraleiter-Bänder (HTS) verwendet und auf der Grundlage der Anforderungen des Antriebsstrangs entwickelt wurde. Die Modellierung des RSFCL mit realitätsnahen Parametern unter adiabatischen und nicht-adiabatischen Bedingungen wird diskutiert. Die Simulationsergebnisse zeigen den effektiven Beitrag der RSFCL zur Begrenzung des Fehlerstroms. Darüber hinaus wird während des Kurzschlusses kein Unterschied zwischen diesen Bedingungen festgestellt, während das nicht-adiabatische Modell nach dem Fehler vorhersagen kann, wie lange es dauert, bis das Kühlmittel die in den Bändern erzeugte zusätzliche Wärme absorbiert und sie in den supraleitenden Zustand zurückführt.

Um die Energie vom Brennstoffzellenstapel effizient zum Motor zu übertragen, werden supraleitende DC- und AC-Kabel integriert. Das spezifische Design dieser Kabel wird zur Verfügung gestellt, und sie werden unter verschiedenen Szenarien durch verschiedene Modelle simuliert, darunter Modelle mit pauschalen Parametern sowie eindimensionale (1-D) und zweidimensionale (2-D) Modelle auf Basis der Finiten-Differenzen-Methode (FDM). Der Anstieg der Kühlmitteltemperatur entlang der Kabellänge kann nru durch das 2-D-Modell berechnet werden. Allerdings ist dieses Modell komplizierter als andere Modelle und erfordert einen höheren Rechenaufwand. Die Simulationsergebnisse zeigen, dass das Modell mit pauschalen Parametern während eines Kurzschlusses immer noch relativ ähnliche Ergebnisse wie das 2-D-Modell als Referenz liefert, und das 1-D-Modell ist die ideale Wahl für die Simulation von Kabeln kurzer Länge, da der Temperaturanstieg über die Länge vernachlässigbar ist.

Der supraleitende Motor für den Antrieb des Propellers wird durch einen DC/AC-Wechselrichter gesteuert. In dieser Arbeit wird ein Permanentmagnet-Synchronmotor (PMSM) betrachtet, dessen Statorwiderstand aufgrund der supraleitenden Spulen sehr gering ist. Der Modellierungsansatz für diesen Motor über ein grundlegendes elektrisches

Ersatzschaltbild wird vorgestellt, und der Regelungsansatz unter Verwendung der feldorientierten Regelungstechnik (FOC), der bekanntesten PMSM-Regelungsmethode, wird erläutert. Zur Simulation des DC/AC-Wechselrichters wird sein Durchschnittsmodell betrachtet und ein Fehlerschutzalgorithmus für den Wechselrichter entwickelt.

Schließlich werden die Ergebnisse der gesamten Antriebsstrangsimulation in einem Simulationsflug-Szenario, das einem realen Flugprofil nachempfunden ist, vorgestellt und diskutiert. Der integrierte Antriebsstrang in SIMULINK wird auch bei Gleich- und Wechselstromkurzschlüssen simuliert, was Einblicke in den Beitrag der einzelnen Komponenten zu den Fehlerereignissen ermöglicht. Die Simulationsergebnisse auf Systemebene bestätigen die Auslegung der einzelnen Komponenten auf der Grundlage der Anforderungen des Antriebsstrangs, die effiziente Wirkung der supraleitenden Komponenten und die Wirksamkeit der Motorantriebssteuerung. Die Kurzschlussanalyse gibt Aufschluss über die Dimensionierung der Komponenten und den Beitrag der RSFCL zum Schutz der Kabel vor Fehlern. Diese Studie zeigt auch die Blockierung des Umrichters durch den Fehlermanagement-Algorithmus, die dazu führt, dass der Motor aufgrund der Motorträgheit zu einem Generator wird, der den Fehler speist. Da die supraleitenden Spulen im Motor einen geringen Widerstand haben, wird vom Motor ein erheblicher Strom erzeugt. Dieser Strom kann als Input für die Auslegung und weitere Analyse des Motors mit fortschrittlicheren Methoden verwendet werden. Insgesamt sind alle modellierten Komponenten durch den in dieser Arbeit erläuterten Ansatz an die Anforderungen anpassbar, was eine vorteilhafte Möglichkeit bietet, sie in breiteren Energiesystemanwendungen einzusetzen.

Abstract

Aviation is responsible for around 2.4% of the annual global carbon emissions. Without further measure, with a yearly passenger traffic growth of 3.6-3.8%, CO2 emissions are expected to double by 2050. This substantial impact has been addressed in recent decades, and multiple projects like EU Flightpath 2050 have been launched to tackle this challenge for carbon-free air travel. Various alternatives to conventional aviation fuel have been investigated to fulfill this objective. The most feasible solutions are sustainable aviation fuel (SAF) and hydrogen. Despite e-fuel, which can only be burnt in gas turbines, green hydrogen can be used in combustion engines or to produce electricity through fuel cell stacks in an electric aircraft. Liquid hydrogen, LH_2, has nearly three times more energy content per unit mass than jet fuel, which makes it an ideal choice for aviation decarbonization. Several studies have identified LH_2 as the long-term solution for long-range flights, replacing sustainable aviation fuel. Using hydrogen in a fuel cell-based electric aircraft has multiple benefits over burning it in gas turbines, including higher efficiency, quieter operation, and elimination of emissions such as NO_x and SO_2 in addition to CO_2. Therefore, several companies like Airbus and Boeing have already considered aviation electrification in different configurations, including fully electric, hybrid electric, and turboelectric aircraft. Due to all the benefits, this work focuses on a fuel cell-based fully electric aircraft.

With the increased required power, the components size and weight increase substantially, especially the hydrogen tank size. Hence, the powertrain components size and weight must be lighter and more efficient, as higher efficiency leads to lower hydrogen consumption and, consequently, smaller tanks. Superconducting material stands as a promising solution, allowing for power transmission with reduced weight and space and more efficiency. This has been a motivation for considering this technology in several projects in aviation electrification, like Airbus's ASCEND project. Thus, an electric aircraft with a superconducting powertrain provides a unique opportunity for a rapid transition to carbon-free aviation. Such a powertrain includes several components, such as the fuel cell stack, superconducting fault current limiter, DC/AC cables and motor, and DC/AC inverter. With all the benefits, the accurate modeling of the entire powertrain and each component is an important issue that must be addressed. This work presents the approach to model such a powertrain in MATLAB/SIMULINK for system simulation in various flight scenarios.

Fuel cell stacks as power source converts hydrogen chemical energy to electric energy. Proton exchange membrane fuel cells (PEMFC) offer suitable properties for aviation applications, hence, this fuel cell type is considered in this work. Fuel cell delivers the power with a non-linear voltage-current variation, meaning that the higher the current, the lower the output voltage. Most research works propose a resistive-capacitive transient behavior for the fuel cell. However, the experiments often show a resistive-inductive behavior or, depending on the fuel cell, a combination of both. In this work, the approach to model any behavior is presented.

To prevent the short circuit currents in the powertrain, a resistive superconducting fault current limiter (RSFCL), using high temperature superconducting (HTS) tapes and designed based on the powertrain requirements, is integrated. The lumped-parameter modeling of the RSFCL in adiabatic and non-adiabatic conditions is discussed. The simulation results show the RSFCL effective contribution in limiting the fault current. Moreover, no difference between these conditions during short circuits is observed, while after fault, the non-adiabatic model can predict how long it takes for the coolant to absorb the additional heat generated in the tapes and return them to the superconducting state.

The superconducting DC and AC cables are integrated to efficiently transfer the energy from the fuel cell stack to the motor. The specific design of these cables is provided, and they are simulated under different scenarios through different models, including lumped-parameter, and one-dimensional (1-D) and two-dimensional (2-D), finite-difference method (FDM) based models. The 2-D model is the only one that can calculate the coolant temperature along the cable length. However, this model is more complicated than other models and has higher computational efforts. The simulation results show that during a short circuit, the lumped-parameter model can still give relatively similar results to the 2-D model as a reference, and the 1-D model is an ideal choice for the simulation of the short-length cables, as temperature rise along the length is negligible.

A superconducting motor for the aircraft propeller, which is controlled by the DC/AC inverter, is integrated. This work considers a permanent magnet synchronous motor (PMSM) with significantly small stator resistance due to the superconducting coils. The modeling approach for this motor via a basic electrical equivalent circuit is presented, and the control approach using the field-oriented control (FOC) technique, as the most well-known PMSM control method, is explained. To simulate the DC/AC inverter, its average model is considered, and a fault protection algorithm for the inverter is developed.

Finally, the results of the entire powertrain simulation in a simulation flight scenario inspired by a real flight profile are presented and discussed. The integrated powertrain in SIMULINK is also simulated under DC and AC short circuits, providing insights into the contribution of each component in the fault events. The system-level simulation results confirm the design of each component based on the powertrain requirements, lossless impact of the superconducting components and validate the effectiveness of the motor drive control. Short circuit analysis indicates an estimation of the components sizing and the RSFCL contribution in protecting the cables from fault. This study also shows the converter blockage by the fault management algorithm, which results in turning the motor into a generator due to motor inertia, feeding the fault. Since superconducting coils in the motor have a small resistance, the motor produces a substantial current, which might be dangerous. This current can be used as an input for the design and further analysis of the motor with more advanced methods. Overall, by following the approach explained in this work, all the modeled components are adjustable with requirements, which provides a beneficial opportunity to use them in wider power system applications.

Table of Contents

List of Figures

List of Tables

List of Acronyms

CO_2	Carbon Dioxide
SAF	Sustainable Aviation Fuel
Mt	Megatonne
DAC	Direct Air Capture
ICE	Internal Combustion Engines
NO_x	Nitrogen Oxides
SO_2	Sulfur Dioxide
MEA	More Electric Aircraft
eVTOL	Electric Vertical Takeoff and Landing
rpm	round per minute
FC	Fuel Cell
PEMFC	Proton Exchange Membrane Fuel Cell
AFC	Alkaline Fuel Cell
PAFC	Phosphoric Acid Fuel Cell
MCFC	Molten Carbonate Fuel Cell
SOFC	Solid Oxide Fuel Cell
DMFC	Direct Methanol Fuel Cell
NASA	National Aeronautics and Space Administration
H_2	Hydrogen
O_2	Oxygen
R-C	Resistive-Capacitive
R-L	Resistive-Inductive
RSFCL	Resistive Superconducting Fault Current Limiter
LTS	Lowe Temperature Superconductor

HTS High Temperature Superconductor

REBCO Rare-Earth Barium Copper Oxide

LN_2 Liquid Nitrogen

LH_2 Liquid Hydrogen

YBCO Yttrium Barium Copper Oxide

PPLP Polypropylene Laminated Paper

1-D One-Dimensional

2-D Two-Dimensional

ADI Alternating Direction Implicit

PMSM Permanent Magnet Synchronous Motor

FEM Finite Element Method

EMF Electromotive Force

PM Permanent Magnet

KVL Kirchhoff's Voltage Law

IGBT Insulated-Gate Bipolar Transistor

MOSFET Metal Oxide Semiconductor Field Effect Transistor

Si Silicon

SiC Silicon Carbide

GaN Gallium Nitride

FOC Field-Oriented Control

List of Symbols

Symbol	Description	Unit
P	Power	W
n	Speed	rpm
I	Current (Instant Amplitude)	A
i	Current (Time-variable)	A
U	Voltage (Instant Amplitude)	V
u	Voltage (Time-variable)	V
t	Time	s
T	Temperature	K
E	Potential	V
l	Length (Thickness)	m
A	Area	m^2
p	Pressure	atm
J	Current Density	A/m^2
c	Concentration	-
R_g	Gas Constant	8.312 J/(mol.k)
F	Faraday's Constant	96485 C/mol
R	Resistance	Ω
L	Inductance	H
C	Capacitance	F
ρ	Resistivity	Ω.m
τ	Time Constant	s
J_c	Critical Current Density	A/m^2
E	Electric Field	V/m

E_c	Critical Electric Field	V/m
T_c	Critical Temperature	K
n	Superconductor's n-Index	-
d	Depth (Thickness)	m
w	Width	m
h_c	Convective Heat Coeffcient	$W/K.m^2$
μ_0	Free Space Permeability	$4\pi\times10^{-7}$ T.m/A
m	Mass	kg
$\dot{m}$	Mass Flow Rate	kg/s
Re	Reynolds Number	-
c_p	Specific Heat capacity	J/(kg.K)
η	Dynamic Viscosity	Pa.s
λ	Fluid Pressure Friction Coefficient	-
q_{rad}	Radiated Heat Input per Area	W/m^2
x	Axial Distance (Length)	m
r	Radius	m
g	Generated Power per Volume	W/m^3
k	Thermal Conductivity	W/(m.K)
Pr	Prandtl Number	-
Nu	Nusselt Number	-
φ	Flux Linkage	Wb
T	Torque	N.m
ω	Rotational Speed	rad/s
J	Inertia	$kg.m^2$
B	Damping Coefficient	N.m/(rad/s)
S	Apparent Power	V.A

P	Active Power	W
Q	Reactive Power	V.AR
E	Energy	J or Wh

1 Introduction

1.1 Motivation and Background

The significant impacts of aviation on global warming, especially CO_2 emissions, have been at the center of attention in recent decades [1], [2]. At present, commercial air travel is responsible for around 2.4% of the annual global carbon emissions [3] and approximately 12% of transport sector CO_2 emissions [4], in which the passenger flights share is 85% (785 Mt) [5]. With a yearly passenger traffic growth of 3.6-3.8%, CO_2 emissions are expected to double by 2050 [4], [6].

To tackle this challenge, multiple measures have been taken in recent years, such as Euro Flightpath 2050 by the European Union, which aims at 75% CO_2 emissions by 2050 [7]. To fulfill this objective, various alternatives to conventional aviation fuel have been proposed and investigated [8], [9]. The results show that to achieve the net zero target, the most feasible solutions are sustainable aviation fuel (SAF) by 2025 and a green fuel by 2030 [4]. A series of projects and studies in recent years have identified green hydrogen produced from renewable energy sources as a vital step to decarbonize aviation [4], [10], [11], [12], [13], [14], [15], [16].

There are two types of SAF, and both are produced from carbon by converting it into hydrocarbon fuel: biofuels and electrofuels (e-fuel) [17], [18]. The difference between these two is that the carbon used to produce biofuel comes from organic sources, while e-fuel uses CO_2 from a non-biogenic source, such as Direct Air Capture (DAC) [19]. Even though the majority of the SAF is biofuel (over 99% of SAF in 2021 was biofuel [17], [18]), due to the increasing availability constraints of sustainable biomass, the capacity to produce biofuels is limited. Therefore, e-fuels might be a better solution than biofuels. A simplified chain of e-fuel production from renewable energy sources for use in aircraft combustion engines is depicted in Figure 1.1 [19].

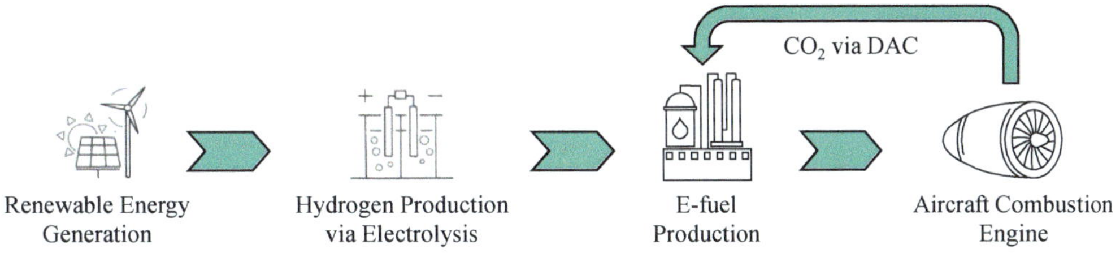

Figure 1.1: Process of e-fuel production and use in the aircraft with combustion engines [19].

Despite e-fuel, which can only be burnt in gas turbines, green hydrogen can be used in both combustion engines or to produce electricity through fuel cell stacks [4], [20], [21]. In the second case, the aircraft will have an electric powertrain. Moreover, liquid hydrogen, LH_2, has nearly three times more gravimetric energy content than jet fuel, which makes it an ideal choice for aviation decarbonization, however, it has lower volumetric energy density, which is a challenge [22]. Several studies, such as [23], [24], [25], have identified LH_2 as the long-term solution for long-range flights, replacing SAF. The processes of hydrogen production from renewable energy sources for use in combustion engines and electric aircraft are respectively illustrated in Figure 1.2(a) and Figure 1.2(b) [21], [26].

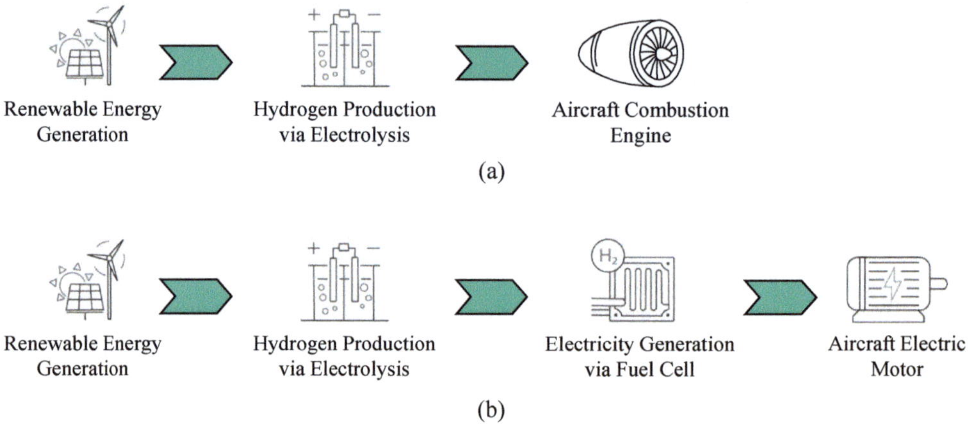

Figure 1.2: Process of green hydrogen production and use in: (a) aircraft with combustion engines, (b) electric aircraft [21], [26].

Considering hydrogen as a propulsion fuel is not a recently-developed concept and was examined in several projects starting from the 18th century, such as Suntan in the U.S. in 1956, CRYOPLANE in Europe in 2000, and HyShot in Australia in 2001 [20]. Even though hydrogen spark-ignited internal combustion engines (ICE) are carbon-neutral in terms of CO_2 emissions, they still produce nitrogen oxides (NO_x) and sulphur dioxide (SO_2), which have negative environmental impacts [20], [27], [28], [29]. These engines deliver lower mechanical power relative to their space and weight than fuel cells and are louder [29], [30]. Aircraft propulsion electrification and using an electric motor on board provide an opportunity to optimize the engine performance and entire powertrain [31]. The UK government-funded Aerospace Technology Institute's FlyZero project concluded that fuel cell-based aircraft offers the most feasible solution for regional flights, while for long-range medium-sized applications, hydrogen-based internal combustion engines are recommended as a practical alternative [22], [32]. A comprehensive comparison between these configurations, including factors like maturity and reliability, falls beyond the scope of this work and needs a separate analysis, and both solutions can be suitable depending on the application. This work focuses on the electric aircraft concept powered by hydrogen through fuel cell stack energy conversion.

1.2 State of the Art on Electric Aircraft

An electric aircraft historically was first built using nickel-cadmium (Ni-Cd) batteries on board in 1972 [31]. With advancements in electrical component technology, the concept of electric systems in aircraft was re-introduced in the late 1980s [31]. This led to the introduction of the "Fly by Wire" concept, which replaced traditional pneumatic and hydraulic systems with electrical devices, as found in the Airbus A320 and later the Boeing 777 [31]. Through a step-by-step approach, electrical components gained broader recognition in aviation, with applications expanding to flight control, engine generators, fuel pumps, and more. This progression led to the concept of the More Electric Aircraft (MEA), later used in aircraft like the Airbus A380 and Boeing 787 [33]. In recent decades, since 2016, there have been over 300 projects and 200 startups globally working on the concept of electric aircraft [34].

There are three main configurations for electrifying the propulsion system of an electric aircraft [31]:

- **Fully Electric:** In this configuration, the powertrain is completely based on the electrical components, which are supplied by batteries or hydrogen through fuel cell stacks. Most of the examples built so far in this structure are low-range, 5-7 passenger aircraft that use batteries on board, like NASA's SCEPTOR X-57 [31], or in electric vertical takeoff and landing (eVTOL) aircraft which are manufactured by several companies, such as Airbus, Lilium, Volocopter, and Aurora [35]. With the increase in the size and number of passengers, consequently higher power, hydrogen would be a more favorable solution as its liquid form provides significantly higher energy density and passenger-range performance compared to batteries [36].
- **Turboelectric:** A turboelectric configuration uses fuel to convert chemical energy into electrical power, either fully or partially, to drive the propulsion. While it has lower efficiency due to conversion losses, it supports novel concepts like distributed propulsion and boundary layer ingestion [31]. The system-level benefits and challenges of this configuration were studied in multiple projects, such as NASA's N3-X [37].
- **Hybrid Electric:** In this configuration, the aircraft is supplied by more than one type of energy source, such as hydrogen-battery-powered aircraft. Hybrid electric configurations can be either series or parallel. In this setup, a temporary energy source like battery energy storage systems can be used during takeoff as an auxiliary source to the main energy source, like hydrogen [38]. With fewer components, this configuration benefits from weight savings. However, the mechanical coupling adds operational and control complexity [31], [39]. This architecture was used in several projects, such as Bauhaus Luftfahrt (BHL) [38] and Rolls-Royce North America (RRNA) Electrically Variable Engine (EVE) [40].

The increased power requirements to build a regional electric aircraft, such as one comparable in size to an Airbus A320, lead to a substantial increase in component size and weight. Superconducting materials offer a promising solution, allowing for the transmission of the same power with reduced weight, compactness, increased efficiency, and almost low electric loss, providing a unique opportunity for aviation electrification [41], [42], [43], [44], [45], [46], [47], [48]. The higher the total power of the electric aircraft, the more interesting it is to use superconducting devices [45]. Furthermore, the dual use of LH_2 on board as an energy carrier and coolant for superconducting tapes can be a significant benefit, as less space will be needed. This has been a motivation to bring superconducting technology on board a fully electric aircraft. A recent example is Airbus's Advanced Superconducting and Cryogenic Experimental powertraiN Demonstrator (ASCEND) project [49]. Considering all these potentials, a model to simulate the superconducting powertrain of an electric aircraft is pivotal. This work focuses on modeling such a system in MATLAB/SIMULINK.

1.3 Architecture and Configuration of the Powertrain for Electric Aircraft

A potential architecture of the superconducting powertrain, which is considered in this work, is shown in Figure 1.3 [49], [50]. This architecture consists of multiple components, starting with the fuel cell as the energy source with a DC voltage output and extending to the propeller's three-phase AC motor. On the DC side of this powertrain, a resistive superconducting fault current limiter (RSFCL) considered to prevent sudden rises in current by potential short circuits is connected to the fuel cell system, and through a bipolar coaxial superconducting DC cable, the power is transported efficiently to other components. The connection between the DC and AC sides is made via the DC/AC inverter, which is used to control the motor speed and torque. Moreover, a superconducting AC cable delivers the power from the inverter to the motor. In this work, liquid nitrogen (LN_2) at a temperature of 77 K is the cooling medium for cryogenically-cooled components. Nevertheless, the models can be easily adapted to other coolants at different temperature levels.

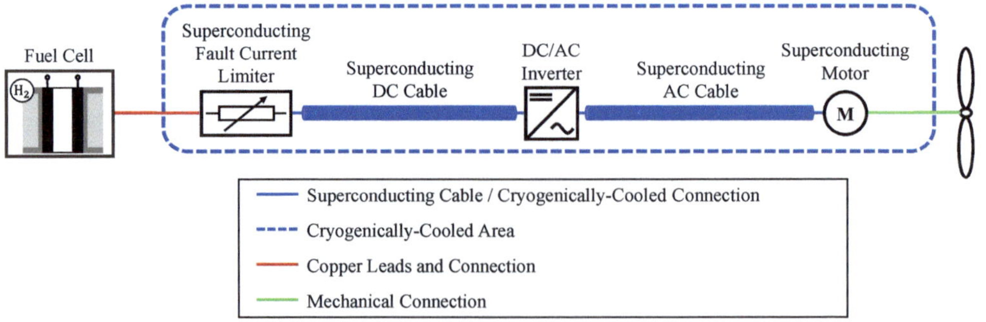

Figure 1.3: Potential architecture of the superconducting powertrain for electric aircraft [49].

In this powertrain, the fuel cell stacks are connected via copper leads at room temperature to the RSFCL, as shown in red in Figure 1.3. The other components are connected via superconducting cables or current leads at cryogenic temperature (77 K). This area is highlighted with a blue dashed line, and the connections are shown in blue in Figure 1.3. Finally, the mechanical connection between the motor and the propeller is demonstrated in green.

Table 1.1 describes the main parameters of this powertrain. This data is used to design and size the components [49], [50].

Table 1.1: Main specifications considered for the electric aircraft powertrain.

Parameter	Symbol	Value
Maximum Power	P_{max}	456 kW
Maximum Speed	n_{max}	1043 rpm
Maximum Nominal AC Current (Peak)	$I_{\mathrm{AC,nom}}$	1.8 kA
Minimum Nominal Pole-to-Pole DC Link Voltage	$U_{\mathrm{DC,nom}}$	300 V
Maximum Nominal DC Current	$I_{\mathrm{DC,nom}}$	1.52 kA
Coolant (LN_2) Temperature	$T_{\mathrm{LN_2}}$	77 K

In this work, all the components are modeled as standalone, and finally, the entire powertrain, with reference to flight profiles and in the event of short circuits, are simulated. Different models are proposed for each component, and different scenarios are simulated to verify their performance. Finally, the complete powertrain simulation results provide insights into the performance of each component and design ideas.

2 Fuel Cell

2.1 Background

The fuel cell is an electrochemical system converting chemical energy into electrical energy [51], [52]. A fuel cell, in general, consists of two electrodes, cathode and anode, and an electrolyte membrane in the middle [51], [52], [53]. Depending on the electrolyte material and chemical reactions inside the fuel cell, there are several types of fuel cell systems currently under investigation in the industry [53], [54], [55]:

- Proton Exchange Membrane Fuel Cell or Polymer Electrolyte Membrane Fuel Cell (PEMFC)
- Alkaline Fuel Cell (AFC)
- Phosphoric Acid Fuel Cell (PAFC)
- Molten Carbonate Fuel Cell (MCFC)
- Solid Oxide Fuel Cell (SOFC)
- Direct Methanol Fuel Cell (DMFC)

A comparison between different fuel cell types is described in Table 2.1 [51], [53], [55], [56], [57], [58].

Table 2.1: Comparison of different fuel cell types [51], [53], [55], [56], [57], [58].

Fuel Cell	Operating Temperature (°C)	Power (kW)	Efficiency (%)	Application
Proton Exchange Membrane Fuel Cell (PEMFC)	60-110	0.01-250	50-70	Cars, Buses, Trucks, Mobility
Alkaline Fuel Cell (AFC)	70-130	0.1-50	50-70	Backup Generators, Primary Power Generators
Phosphoric Acid Fuel Cell (PAFC)	175-210	50-1000	40-45	Buildings, Hotels, Hospitals, Utilities
Molten Carbonate Fuel Cell (MCFC)	550-650	200-100,000	50-60	Utilities, Large Scale Power Generation
Solid Oxide Fuel Cell (SOFC)	500-1000	0.5-2000	40-72	Medium to Large Scale Power Generation
Direct Methanol Fuel Cell (DMFC)	70-130	0.001-100	40	Mobility

Among all fuel cell types, PEMFCs exhibit properties suitable for aeronautic applications. These appealing features include silent operation, fast startup and response to alternating loads, high efficiency combined with high power density, solid electrolyte, lightweight, and low operating temperature range [59], [60], [61]. Given these advantages, many companies such as NASA, Boeing, and Airbus have been studying and considering the use of PEMFCs in aircraft [59], [60], [61], [62], [63]. These are the motivations for considering and studying PEMFCs for an electric aircraft powertrain in this work.

Proton exchange membrane fuel cells are constructed using polymer electrolyte membranes (mostly Nafion®) [64]. Figure 2.1 demonstrates the electrochemical processes occurring in the PEMFC, including chemical reactions in anode, cathode, and membrane electrolyte [51]. In this fuel cell, hydrogen is fed to the anode where it splits into ions (protons) and electrons because of the anode catalysts. While electrons are forced to flow in an external electrical circuit, generating electric energy to supply a load, the H^+ protons pass through the electrolyte to the cathode. At the cathode, the hydrogen protons and electrons combine with the oxygen, which is typically drawn from the air, producing water (possibly also in the form of vapor) and a small quantity of heat [51], [65]. In Figure 2.1, i_{FC} and u_{FC} are the fuel cell generated current and voltage, respectively.

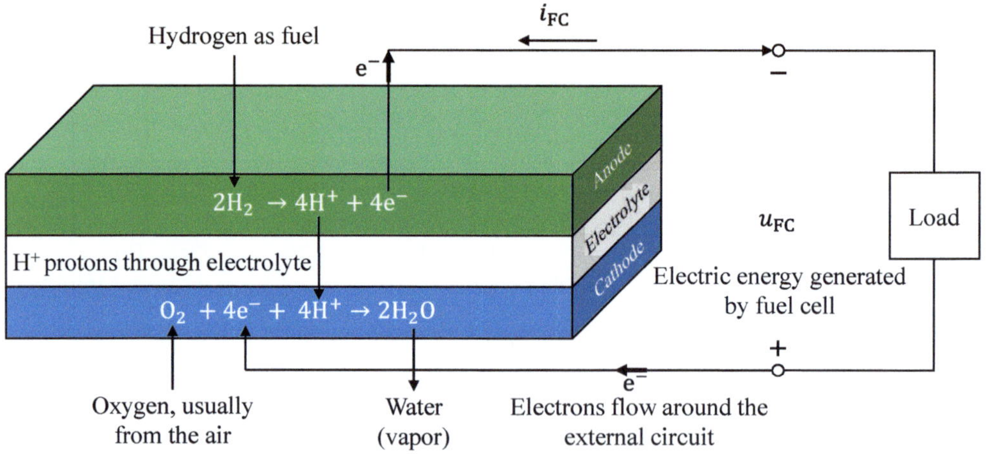

Figure 2.1: Schematic representation of the electrochemical processes in a PEMFC [51].

2.2 Fuel Cell Stack Design

2.2.1 Fuel Cell Unit Properties

In this work, a fuel cell unit described in [66] is considered. The properties and parameters of this fuel cell are given in Table 2.2.

Table 2.2: Fuel cell unit properties [66].

Parameter	Symbol	Value
Membrane Thickness	l_{M}	178 μm
Fuel Cell Active Area	A_{FC}	64 cm^2
Cell Operating Temperature	T	333 K
Hydrogen Pressure	p_{H_2}	1 atm
Oxygen Pressure	p_{O_2}	0.2095 atm
Fuel Cell Maximum Current Density	$J_{\mathrm{FC,max}}$	469 $\mathrm{mA/cm}^2$

2.2.2 Voltage-Current Characteristic

2.2.2.1 Generated voltage (Nernst potential)

In a PEMFC, the thermodynamic potential of a unit cell that represents its voltage is called Nernst potential[1] E_{Nernst}. This voltage is calculated with equation (2.1) [67]. In this equation, T is the cell temperature, and p_{H_2} and p_{O_2} are the hydrogen and oxygen partial pressure. The first term in this equation, 1.229 V, is when the water product is in liquid form; in the case of the gaseous water product (vapor), it is reduced to 1.18 V.

$$E_{\mathrm{Nernst}} = 1.229 - 0.85 \times 10^{-3}(T - 298.15) + 4.31 \times 10^{-5}T \cdot \left[\ln\left(p_{\mathrm{H}_2}\right) + 0.5\ln\left(p_{\mathrm{O}_2}\right)\right] \tag{2.1}$$

As seen in equation (2.1), Nernst potential varies with cell temperature and input hydrogen and oxygen pressure. With an increase in the temperature, E_{Nernst} decreases, while it is boosted with higher oxygen and hydrogen pressures. The variation of Nernst potential in the temperature range of 290-370 K at different hydrogen and oxygen pressures is depicted in Figure 2.2. In this figure, the red curve shows E_{Nernst} when the pressures of hydrogen and oxygen are 1 atm and 0.2095 atm, while in the case of the blue curve, they are fed at higher pressures of 1.5 atm and 1 atm, respectively.

[1] Named after Walther Nernst, a German physical chemist who formulated the equation.

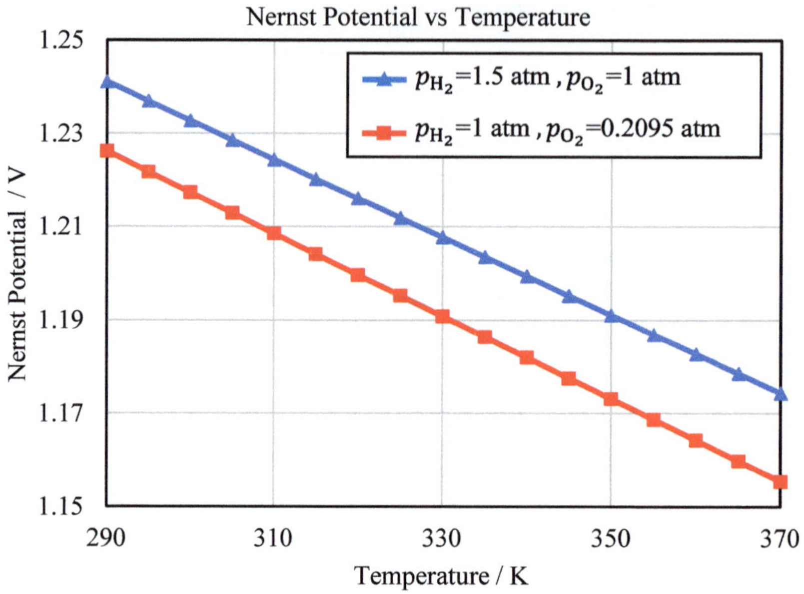

Figure 2.2: Nernst potential variation with temperature, and hydrogen and oxygen partial pressure.

Nernst potential is an ideal voltage a PEMFC can generate, but the output voltage is lower due to the different voltage drops inside the fuel cell unit. These voltage drops are [51], [67]:

- Activation voltage drop (u_{act})
- Concentration voltage drop (u_{con}) or mass transport voltage loss
- Ohmic voltage drop (u_{Ω})

2.2.2.2 Activation voltage drop

Activation voltage drop represents the voltage loss due to the anode and cathode activation to initiate a chemical reaction between reactants that transfers the electrons to or from the electrode [66]. This loss is caused by the lack of electro-catalytic activity and slowness of the reactions occurring at the surface of these two electrodes [51], [66]. Equation (2.2) describes the calculation of the activation voltage drop [51].

$$u_{\mathrm{act}} = -\left[\xi_1 + \xi_2 \cdot T + \xi_3 \cdot T \cdot \ln\left(c_{O_2}\right) + \xi_4 \cdot T \cdot \ln(i_{\mathrm{FC}})\right] \tag{2.2}$$

In this equation, c_{O_2} is the oxygen concentration which is calculated with equation (2.3) from the oxygen partial pressure p_{O_2} and the cell temperature T.

$$c_{O_2} = \frac{p_{O_2}}{5.08 \times 10^6 e^{-\left(\frac{498}{T}\right)}} \tag{2.3}$$

Moreover, ξ_i $(i = 1 - 4)$ are the parametric coefficients to describe the cell model. These parameters are shown in equation (2.4).

$$\begin{aligned} &\xi_1 = -0.948, \xi_3 = 7.6 \times 10^{-5}, \xi_4 = -1.93 \times 10^{-4} \\ &\xi_2 = 2.86 \times 10^{-3} + 2 \times 10^{-4} \ln(A_{\mathrm{FC}}) + 4.3 \times 10^{-5} \ln(c_{\mathrm{H_2}}) \end{aligned} \tag{2.4}$$

In equation (2.4), A_{FC} is the fuel cell active area, and $c_{\mathrm{H_2}}$ is the hydrogen concentration calculated with equation (2.5) from the input hydrogen partial pressure $p_{\mathrm{H_2}}$ and the operating temperature T.

$$c_{\mathrm{H_2}} = \frac{p_{\mathrm{H_2}}}{1.09 \times 10^6 e^{\left(\frac{77}{T}\right)}} \tag{2.5}$$

As seen above, the equation (2.2) is not able to calculate the activation voltage drop at no load condition named as $u_{\mathrm{act,NL}}$. This value must be given manually with a compromise and estimation. In this work, this parameter is considered 0.2 V.

2.2.2.3 Concentration voltage drop

Concentration voltage loss results from the change in the reactants concentration at the surface of the electrodes as fuel is consumed [66]. Since this loss is associated with the failure to transport efficient reactants to the anode and cathode surface, it is often called mass transport loss [51]. The concentration voltage drop u_{con} varies with the temperature T and fuel cell generated current density J_{FC} and is calculated with equation (2.6) [52].

$$u_{\mathrm{con}} = -\frac{R_{\mathrm{g}} \cdot T}{2F} \cdot \ln\left(1 - \frac{J_{\mathrm{FC}}}{J_{\mathrm{FC,max}}}\right) \tag{2.6}$$

In equation (2.6), R_{g} is the gas constant (8.314 J/mol.K), the value 2 is the number of exchange protons per mole of reactant (in the case of the PEMFC), and F is Faraday's constant (96485 C/mol). It is worthwhile to mention that the fuel cell current density J_{FC} is calculated with equation (2.7) from fuel cell current i_{FC} and active area A_{FC}.

$$J_{\mathrm{FC}} = \frac{i_{\mathrm{FC}}}{A_{\mathrm{FC}}} \tag{2.7}$$

2.2.2.4 Ohmic voltage drop

Ohmic voltage drop or also often called resistive voltage loss is caused by the membrane resistance to the proton flow (named R_{M}) and the electrodes contact and material resistance to the electron flow (named R_{C}) [51]. Fuel cell total equivalent resistance named R_{Ω} is therefore calculated with equation (2.8).

$$R_\Omega = R_\mathrm{M} + R_\mathrm{C} \tag{2.8}$$

Contact resistance R_C is generally constant in a temperature range of 320-370 K [68], but is difficult to predict. In this work, R_C is considered 20 mΩ.cm^2 [66]. On the opposite, the membrane resistance R_M is significantly dependent on the fuel cell temperature T and current density J_FC. Membrane resistivity ρ_M is calculated with equation (2.9) [66].

$$\rho_\mathrm{M} = \frac{181.6\left[1 + 0.03J_\mathrm{FC} + 0.062\left(\frac{T}{303}\right)^2 \cdot J_\mathrm{FC}^{2.5}\right]}{[\psi - 0.634 - 3J_\mathrm{FC}] \cdot e^{4.18\left(\frac{T-303}{T}\right)}} \tag{2.9}$$

In this equation, the term 181.6/(ψ-0.634) is the membrane resistivity at no-load condition (J_FC=0 A/m^2) and a temperature of 303 K (30 °C). The exponential term in the denominator is the temperature correction factor for cell temperatures different from 303 K, and the term in the square bracket in the numerator is to fit the experiment data [68]. Furthermore, the parameter ψ is adjustable depending on the membrane material, ranging from 14 to 23 [66], [68]. Based on [66], [67], [68], this parameter is considered 23 in this work.

With the calculated ρ_M and membrane length l_M and fuel cell active area A_FC, membrane resistance R_M is calculated with equation (2.10).

$$R_\mathrm{M} = \rho_\mathrm{M} \cdot \frac{l_\mathrm{M}}{A_\mathrm{FC}} \tag{2.10}$$

The Ohmic voltage drop u_Ω is finally calculated with equation (2.11) from fuel cell generated current i_FC and the total fuel cell equivalent resistance R_Ω [66].

$$u_\Omega = R_\Omega \cdot i_\mathrm{FC} \tag{2.11}$$

Knowing the generated voltage and all the voltage drops in a fuel cell unit, its final output voltage u_FC is calculated with equation (2.12).

$$u_\mathrm{FC} = E_\mathrm{Nernst} - u_\mathrm{act} - u_\mathrm{con} - u_\Omega \tag{2.12}$$

2.2.2.5 Polarization Curve

A fuel cell polarization curve describes the non-linear voltage-current behavior of a fuel cell unit. Figure 2.3 depicts the polarization curve of the PEMFC unit described above. This figure shows the variation of fuel cell output voltage with current; the higher the output current, the lower the voltage. Moreover, this figure demonstrates the significant share of activation voltage drop compared to concentration and Ohmic voltage losses.

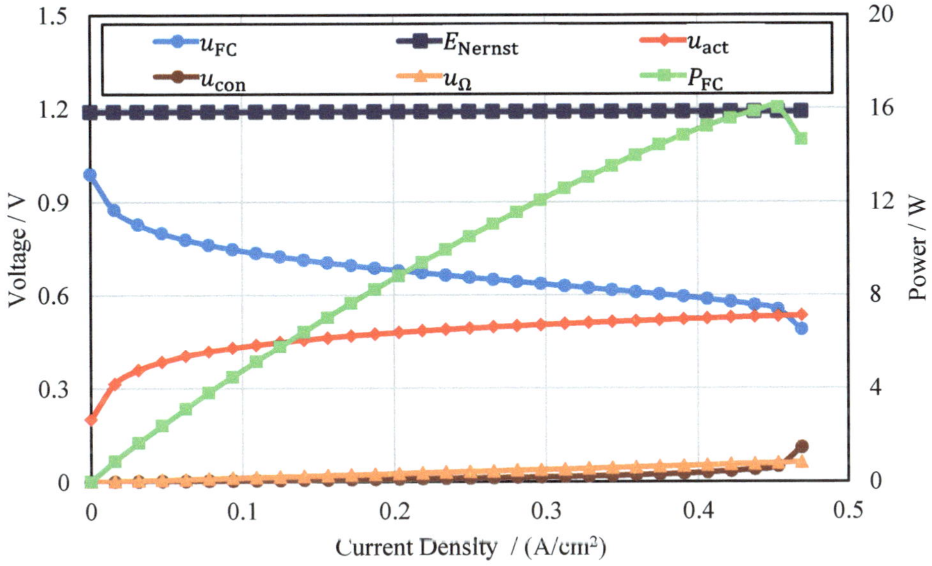

Figure 2.3: Fuel cell unit polarization curve including the different voltage drops (calculated with data from Table 2.2).

In Figure 2.3, the fuel cell power P_{FC} which is calculated from the output voltage u_{FC} and current i_{FC} with equation (2.13) is also illustrated.

$$P_{FC} = u_{FC} \cdot i_{FC} \tag{2.13}$$

The polarization curve shows that the fuel cell power is increasing with the current up to nearly the fuel cell maximum current density $J_{FC,max}$. However, at current densities higher than $J_{FC,max}$, since concentration voltage drop tends to increase significantly, the output voltage suddenly drops leading to a power reduction. Therefore, to find a compromise between fuel cell voltage, current, and power, the optimal operating condition of a fuel cell is at current densities close but with a margin less than $J_{FC,max}$. For the PEMFC considered in this work, the nominal optimal condition is the following:

$$J_{FC,nom} = 0.375\ \text{A/cm}^2, U_{FC,nom} = 0.6\ \text{V}, P_{FC,nom} = 14.4\ \text{W} \tag{2.14}$$

2.2.3 Stack Design

A fuel cell used in an electrical network is made up of several cells called fuel cell stack. To design a fuel cell stack for the powertrain network described in chapter 1, the nominal stack output voltage $U_{FC,st,nom}$ of 300 V and the power $P_{FC,st,nom}$ of 456 kW equal to the maximum required power P_{max} (See Table 1.1 in chapter 1) are considered. The output voltage of 300 V is the minimum voltage the fuel cell stack generates at its highest power, 456 kW.

If the nominal output voltage of a fuel cell unit is $U_{\mathrm{FC,nom}}$, a fuel cell stack composed of N_{c} number of cells in series can generate a total output voltage of $U_{\mathrm{FC,st,nom}}$ which is calculated with equation (2.15).

$$U_{\mathrm{FC,st,nom}} = U_{\mathrm{FC,nom}} \cdot N_{\mathrm{c}} \tag{2.15}$$

To deliver the maximum requested power $P_{\max}$, the fuel cell stack must generate a nominal current of $I_{\mathrm{FC,st,nom}}$ at its nominal output voltage $U_{\mathrm{FC,st,nom}}$. Since the voltage and power are known, this output current is then calculated with equation (2.16).

$$I_{\mathrm{FC,st,nom}} = \frac{P_{\max}}{U_{\mathrm{FC,st,nom}}} \tag{2.16}$$

Then, since the fuel cell nominal current density $J_{\mathrm{FC,nom}}$ is known, the right amount of the cells active area for the stack named as $A_{\mathrm{FC,st}}$ must be found. This parameter is calculated with equation (2.17).

$$A_{\mathrm{FC,st}} = \frac{I_{\mathrm{FC,st,nom}}}{J_{\mathrm{FC,nom}}} \tag{2.17}$$

Table 2.3 outlines the final properties of the fuel cell stack designed for the electric aircraft powertrain. Additionally, the stack polarization curve is shown in Figure 2.4. In this curve, the stack voltage $u_{\mathrm{FC,st}}$ for N_{c} number of cells in series is calculated from stack total Nernst potential $E_{\mathrm{Nernst,st}}$, activation voltage drop $u_{\mathrm{act,st}}$, concentration voltage drop $u_{\mathrm{con,st}}$ and Ohmic voltage drop $u_{\Omega,\mathrm{st}}$ with equation (2.18). The calculation of these voltages is also shown in equation (2.18) where E_{Nernst}, u_{act}, u_{con} and u_{Ω} represent voltages existing for a single cell. Lastly, the stack power $P_{\mathrm{FC,st}}$ is determined from $u_{\mathrm{FC,st}}$ and $i_{\mathrm{FC,st}}$ using the same method described in equation (2.13).

Table 2.3: Fuel cell stack properties, designed based on the cell unit properties, as described in Table 2.2.

Parameter	Symbol	Value
Number of Cells in Series	N_{c}	500
Fuel Cell Stack Active Area	$A_{\mathrm{FC,st}}$	0.385 m^2
Nominal Stack Output Voltage	$U_{\mathrm{FC,st,nom}}$	300 V
Nominal Stack Output Current	$I_{\mathrm{FC,st,nom}}$	1.52 kA
Nominal Stack Output Power	$P_{\mathrm{FC,st,nom}} = P_{\max}$	456 kW

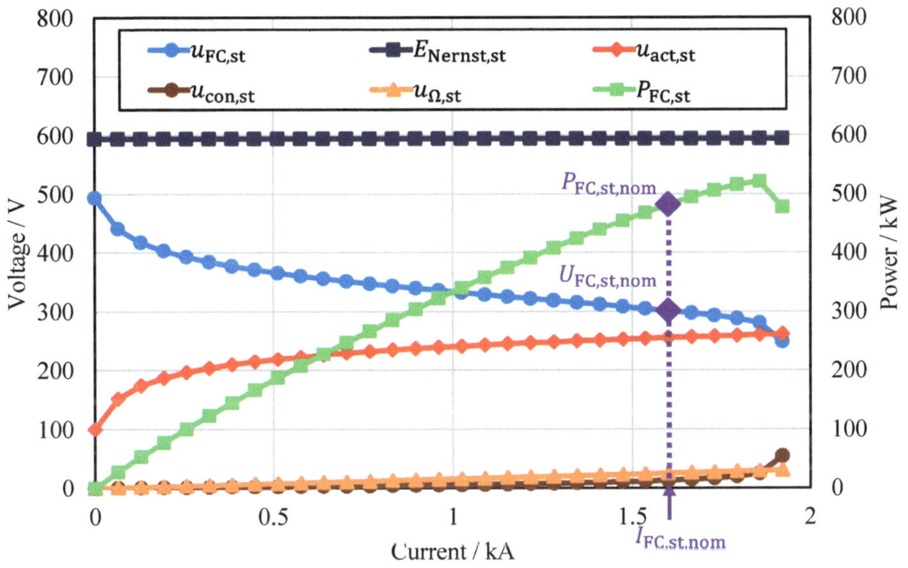

Figure 2.4: Fuel cell stack polarization curve including the different voltage drops (calculated with data from Table 2.2 and Table 2.3).

$$\begin{aligned} E_{\text{Nenrst,st}} &= N_{\text{c}} \cdot E_{\text{Nenrst}} \\ u_{\text{act,st}} &= N_{\text{c}} \cdot u_{\text{act}} \\ u_{\text{con,st}} &= N_{\text{c}} \cdot u_{\text{con}} \\ u_{\Omega,\text{st}} &= N_{\text{c}} \cdot u_{\Omega} \\ u_{\text{FC,st}} = E_{\text{Nernst,st}} - u_{\text{act,st}} - u_{\text{con,st}} &- u_{\Omega,\text{st}} = N_{\text{c}} \cdot u_{\text{FC}} \end{aligned} \tag{2.18}$$

2.3 Fuel Cell, Stationary, Electrochemical Model

An electrochemical equivalent circuit depicted in Figure 2.5 demonstrates the stationary (steady-state) behavior of a fuel cell stack. In this circuit, all the voltages explained in the previous section are represented as controlled (dependent) voltage sources as they are not constant and vary with other parameters like current and temperature. The stack generated voltage is $E_{\text{Nernst,st}}$ and the voltage drops $u_{\text{act,st}}$, $u_{\text{act,st}}$ and $u_{\Omega,\text{st}}$ are shown as sources with opposite polarization.

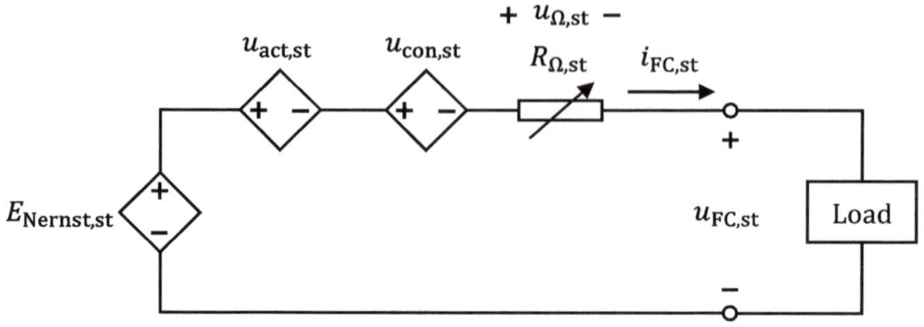

Figure 2.5: Fuel cell stack stationary electrical equivalent circuit.

Since Ohmic losses are proportional to the current, they are represented as a non-linear resistor (due to variability with temperature and current) with a resistance of $R_{\Omega,\mathrm{st}}$. This equivalent resistance of the stack Ohmic losses is calculated with equation (2.19).

$$R_{\Omega,\mathrm{st}} = \frac{u_{\Omega,\mathrm{st}}}{i_{\mathrm{FC,st}}} \tag{2.19}$$

With this equivalent circuit, a load current variation leading into the change in the fuel cell voltages can be monitored. An example is shown in Figure 2.6 when the stack current (load current) varies in 2-seconds steps, causing the step changes of the stack voltage. In this figure, the stack current increases from 0 A to 500 A $t = 2$ s and then reaches its highest value at 1.52 kA at $t = 4$ s. This current then reduces to 800 A at $t = 6$ s and finally is set at 300 A at $t = 8$ s. equal to the nominal stack current, corresponsing to the stack lowest voltage at 300 V.

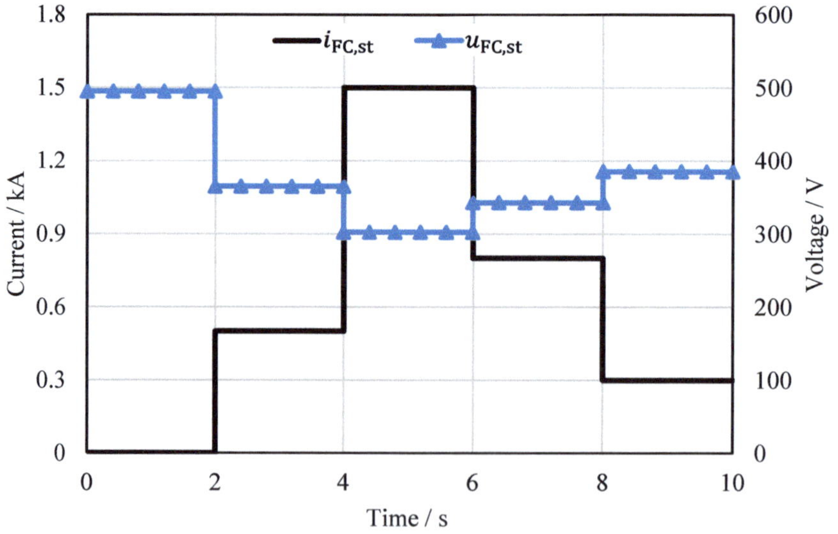

Figure 2.6: Fuel cell stack stationary model results (calculated with data from Table 2.3 and Figure 2.5).

2.4 Fuel Cell, Transient, Electrochemical Model

Because of the fuel cell electrochemical inertia, voltage changes in response to variations in load current do not happen instantaneously. Instead, a transient period is required for the system to reach steady-state conditions. Therefore, it is important to consider not only the stationary behavior of the fuel cell but also its dynamic and transient characteristics. To introduce the transient behavior to the fuel cell, most of the previous works like [67], [69], [70], [71], [72], [73] have proposed a resistive-capacitive equivalent circuit. However, in some works, a resistive-inductive equivalent circuit can be detected [74], [75], [76], [77], [78], [79]. One of the reasons behind the transient resistive-inductive behavior of a fuel cell seems to be related to a phenomenon called fuel starvation or also called reactant starvation which is caused by inadequate fuel supply [77].

2.4.1 Resistive-Capacitive Electrical Equivalent Circuit

Figure 2.7 depicts the resistive-capacitive (R-C) equivalent circuit of a fuel cell stack where $u_{\mathrm{act,st,NL}}$ represents the stack activation voltage drop at no-load condition ($i_{\mathrm{FC,st}}$=0 A). This parameter is calculated with equation (2.20) from the number of cells N_{c} and the activation voltage drop of one cell at no-load condition $u_{\mathrm{act,NL}}$. The parameter $u_{\mathrm{act,st,NL}}$ is 100 V in this work (500 × 0.2 V). Moreover, the parameters R_{d} and C_{d} are respectively dynamic resistance and capacitance with a dynamic voltage of u_{d} to model the fuel cell transient behavior. The stack output voltage $u_{\mathrm{FC,st}}$ is then calculated with equation (2.21).

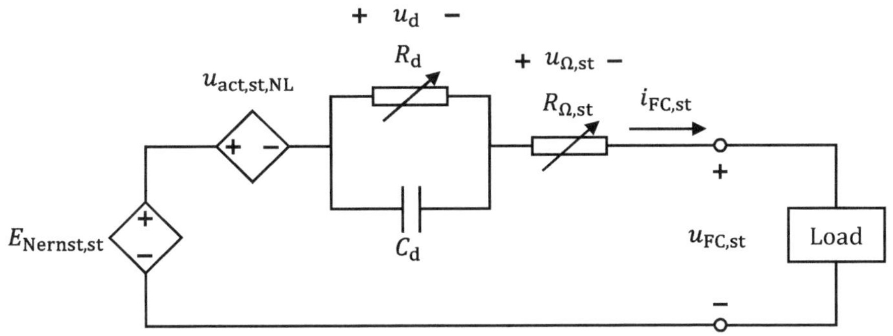

Figure 2.7: Resistive-capacitive (R-C) transient equivalent circuit of the fuel cell stack [67].

$$u_{\mathrm{act,st,NL}} = N_{\mathrm{c}} \cdot u_{\mathrm{act,NL}} \tag{2.20}$$

$$u_{\mathrm{FC,st}} = E_{\mathrm{Nernst,st}} - u_{\mathrm{act,st,NL}} - u_{\mathrm{d}} - u_{\Omega,\mathrm{st}} \tag{2.21}$$

The dynamic voltage u_{d} is a sum of activation and concentration voltage drops deducting $u_{\mathrm{act,st,NL}}$. This is shown in equation (2.22).

$$u_{\mathrm{d}} = u_{\mathrm{act,st}} + u_{\mathrm{con,st}} - u_{\mathrm{act,st,NL}} \tag{2.22}$$

This voltage comprises two terms: transient and steady-state. The steady-state component of u_{d} is characterized by the resistance R_{d}, while the capacitance C_{d} simulates its transient behavior. The dynamic resistance R_{d} is calculated with equation (2.23).

$$R_{\mathrm{d}} = \frac{u_{\mathrm{d}}}{i_{\mathrm{FC,st}}} \tag{2.23}$$

Furthermore, the constant capacitance C_{d} value is determined through an experiment conducted on a chosen fuel. The procedure involves operating the fuel cell under no-load condition (0 A) and then increasing its current at time $t = t^*$ to a specified value, denoted as I. The dynamic voltage at current I denoted by $u_{\mathrm{d,I}}$ for time t before and after t^* is found with equation (2.24), where the exponential term describes the transient behavior, and $R_{\mathrm{d,I}}$ is the corresponding dynamic resistance at the stack current of $i_{\mathrm{FC,st}}=I$ (calculated from equation (2.23) based on equation (2.22)). Moreover, the parameter τ_{RC} is the R-C branch time constant in the equivalent circuit that is calculated with equation (2.25).

$$u_{\mathrm{d,I}} = \begin{cases} 0 & ; t < t^* \\ R_{\mathrm{d,I}} \cdot I \cdot \left(1 - e^{\frac{-(t-t^*)}{\tau_{\mathrm{RC}}}}\right) & ; t \geq t^* \end{cases} \tag{2.24}$$

$$\tau_{\mathrm{RC}} = R_{\mathrm{d,I}} \cdot C_{\mathrm{d}} \tag{2.25}$$

It takes approximately five times the time constant of an R-C circuit for the system to reach its steady-state condition. This duration is referred to as the steady-state time and is denoted by the symbol t_{ss}. The value of the capacitance C_{d} is then calculated from equation (2.25) with equation (2.26).

$$C_{\mathrm{d}} = \frac{\tau_{\mathrm{RC}}}{R_{\mathrm{d,I}}} = \frac{t_{\mathrm{ss}}}{5R_{\mathrm{d,I}}} \tag{2.26}$$

Considering U_{NL} as the stack voltage at no-load condition and U_{ss} as its steady-state voltage at a current of I, the stack voltage $u_{\mathrm{FC,st}}$ including its R-C transient is calculated based on equations (2.21) and (2.24) with equation (2.27).

$$u_{\mathrm{FC,st}} = \begin{cases} U_{\mathrm{NL}} & ; t < t^* \\ U_{\mathrm{ss}} + R_{\mathrm{d,I}} \cdot I \cdot e^{\frac{-(t-t^*)}{\tau_{\mathrm{RC}}}} & ; t \geq t^* \end{cases} \tag{2.27}$$

Having calculated all the parameters, the results of the above-mentioned experiment in the case of the R-C behavior is demonstrated in Figure 2.8. Based on this figure, the required inputs, including I and t_{ss} can be derived, and the equivalent capacitance C_{d} is calculated.

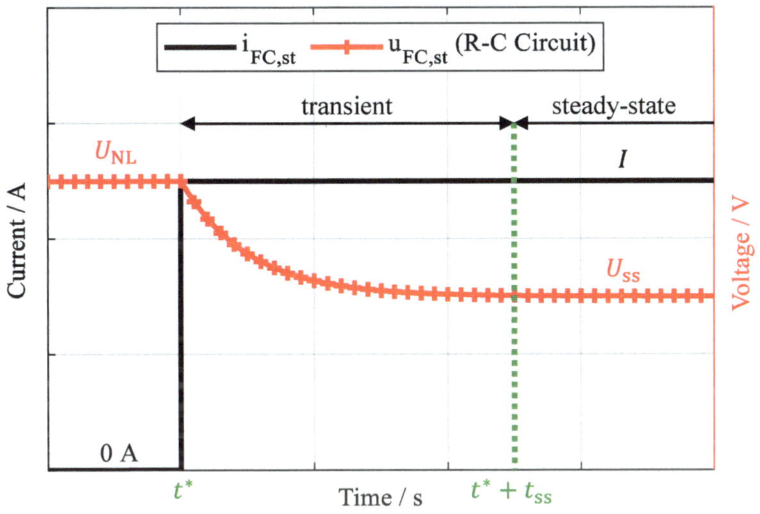

Figure 2.8: Transient voltage variation of a fuel cell stack with a current increase in the case of the resistive-capacitive response (calculated with equation (2.27)).

2.4.2 Resistive-Inductive Electrical Equivalent Circuit

A resistive-inductive (R-L) equivalent circuit for a fuel cell stack can be proposed using the same approach as the R-C equivalent circuit. This R-L circuit is depicted in Figure 2.9, where R_d and L_d represent the equivalent resistance and inductance to model the fuel cell dynamic (transient) behavior. Additionally, the controlled-voltage source with the amplitude of $u_{\mathrm{ss,st}}$ indicates the stack steady-state voltage, subtracting its Ohmic losses, which are already represented in the circuit by its equivalent resistance $R_{\Omega,\mathrm{st}}$. Eventually, the parameter $u_{\mathrm{ss,st}}$ is calculated using equation (2.28), leading to the calculation of the stack output voltage $u_{\mathrm{FC,st}}$ as shown in equation (2.29).

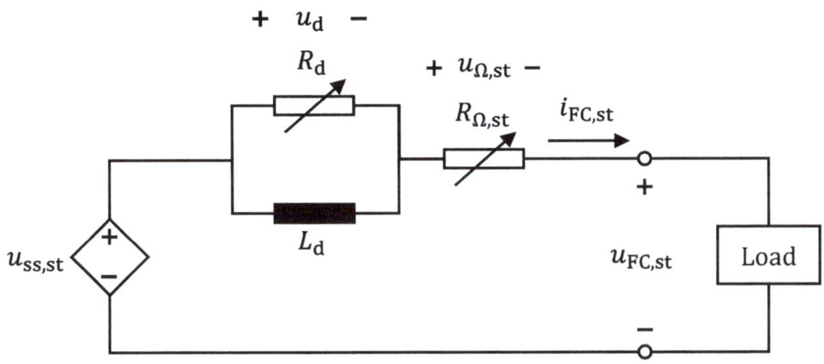

Figure 2.9: Resistive-inductive (R-L) transient equivalent circuit of the fuel cell stack [80].

$$u_{ss,st} = E_{Nernst,st} - u_{act,st} - u_{con,st} \tag{2.28}$$

$$u_{FC,st} = u_{ss,st} - u_{d} - u_{\Omega,st} \tag{2.29}$$

In the R-L circuit, the dynamic voltage u_d is calculated using the same method as R-C circuit described above. This voltage is then determined using equation (2.30) at times t before and after t^*, where τ_{RL} is the R-L branch time constant given in equation (2.31). As observed, the R-L branch does not affect the steady-state voltage. However, during transient conditions, the voltage u_d reaches a maximum value of $R_{d,I} \cdot I$. This causes a sudden drop in the stack output voltage, referred to as the minimum transient voltage and denoted by $U_{tr,min}$. Given the steady-state stack voltage for current I as U_{ss}, the voltage $U_{tr,min}$ is found using equation (2.32). It is noteworthy that, unlike the R-C equivalent circuit, the R-L branch is not related to the internal fuel cell voltages, and it is added solely to simulate its transient behavior.

$$u_d = \begin{cases} 0 & ; t < t^* \\ R_{d,I} \cdot I \cdot e^{\frac{-(t-t^*)}{\tau_{RL}}} & ; t \geq t^* \end{cases} \tag{2.30}$$

$$\tau_{RL} = \frac{L_d}{R_{d,I}} \tag{2.31}$$

$$U_{tr,min} = U_{ss} - R_{d,I} \cdot I \tag{2.32}$$

Since $U_{tr,min}$ and U_{ss} are known, the equivalent dynamic resistance at the current I denoted by $R_{d,I}$, is then calculated with equation (2.33).

$$R_{d,I} = \frac{U_{ss} - U_{tr,min}}{I} \tag{2.33}$$

Similar to the R-C circuit, the time that the R-L branch reaches its steady-state conditions is indicated by t_{ss}, which is about five times the time constant τ_{RL} due to its exponential variations. Therefore, the inductance L_d can be calculated from equation (2.31) using equation (2.34).

$$L_d = \tau_{RL} \cdot R_{d,I} = \frac{t_{ss} \cdot R_{d,I}}{5} \tag{2.34}$$

Finally, the stack voltage $u_{FC,st}$ including its R-L transient is calculated based on equations (2.28), (2.29), and (2.30) using equation (2.35) (U_{NL}: output voltage at no-load condition). Moreover, the variation of this voltage with the previously described experiment should resemble Figure 2.10 based on the theoretical equations explained above.

$$u_{FC,st} = \begin{cases} U_{NL} & ; t < t^* \\ U_{ss} - R_{d,I} \cdot I \cdot e^{\frac{-(t-t^*)}{\tau_{RL}}} & ; t \geq t^* \end{cases} \tag{2.35}$$

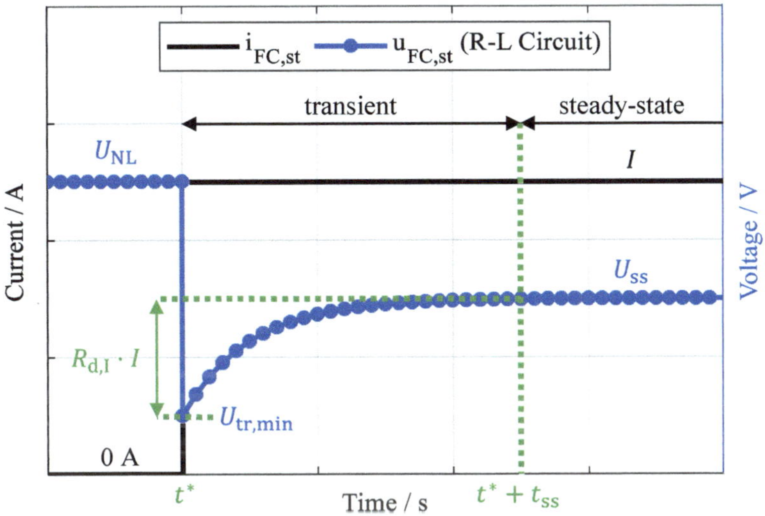

Figure 2.10: Transient voltage variation of a fuel cell stack with a current increase in the case of the resistive-inductive response (calculated with equation (2.35)).

2.4.3 Combination of Resistive-Capacitive and Resistive-Inductive Electrical Equivalent Circuits

A fuel cell system might not exhibit purely resistive-capacitive or resistive-inductive behavior when the load current changes. Instead, its response could include characteristics of both. As indicated in multiple studies, such as [75], [77], an R-L behavior is observed when the current increases, whereas when the current decreases, an R-C response is detected.

2.4.4 Transient Behavior Selection Algorithm

As explained above, a model capable of simulating a fuel cell with a combination of R-C and R-L behaviors, in addition to the models with pure R-C and R-L responses, is necessary. This model is depicted in Figure 2.11, where the parallel R-C and R-L equivalent circuits operate with the stack (load) current $i_{FC,st}$ as input, and respectively, generating the stack voltages $u_{FC,st,RC}$ (equation (2.27)) and $u_{FC,st,RL}$ (equation (2.35)) at the output. An algorithm then selects one of these voltages based on the user's choice of the fuel cell behavior.

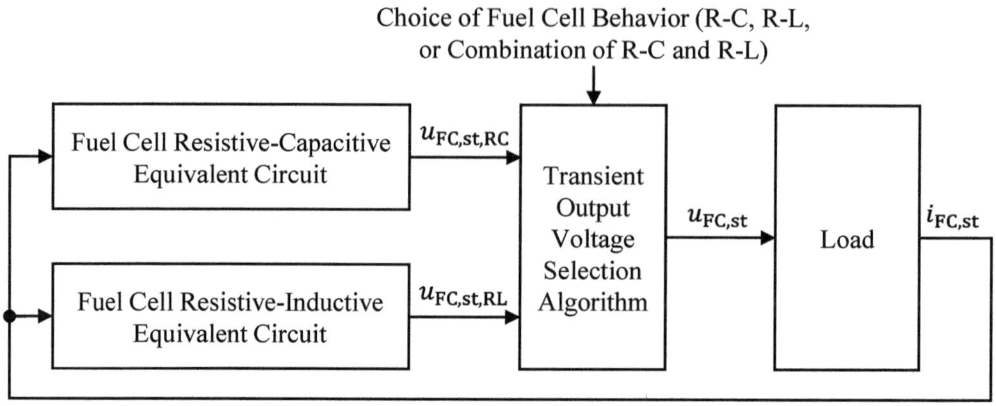

Figure 2.11: Fuel cell model schematic highlighting the process to simulate its transient behavior with different potential responses (R-C, R-L, or a combination of both).

This algorithm is illustrated in Figure 2.12. In this figure, *z* is the number of time steps representing the time loop. A parameter with an index *z*-1 indicates the parameter's value at the previous time step, while index *z* denotes its value at the new (current) time step. As observed in Figure 2.12, in the case of choosing a model with the combination of both R-C and R-L behaviors by the user, the change of current is checked. If the current is increasing, then the voltage generated by the R-L model is selected. If the current decreases, the voltage at the output of the R-C model is selected. And, if the current is constant, the voltage calculated at the previous time step is used with no change.

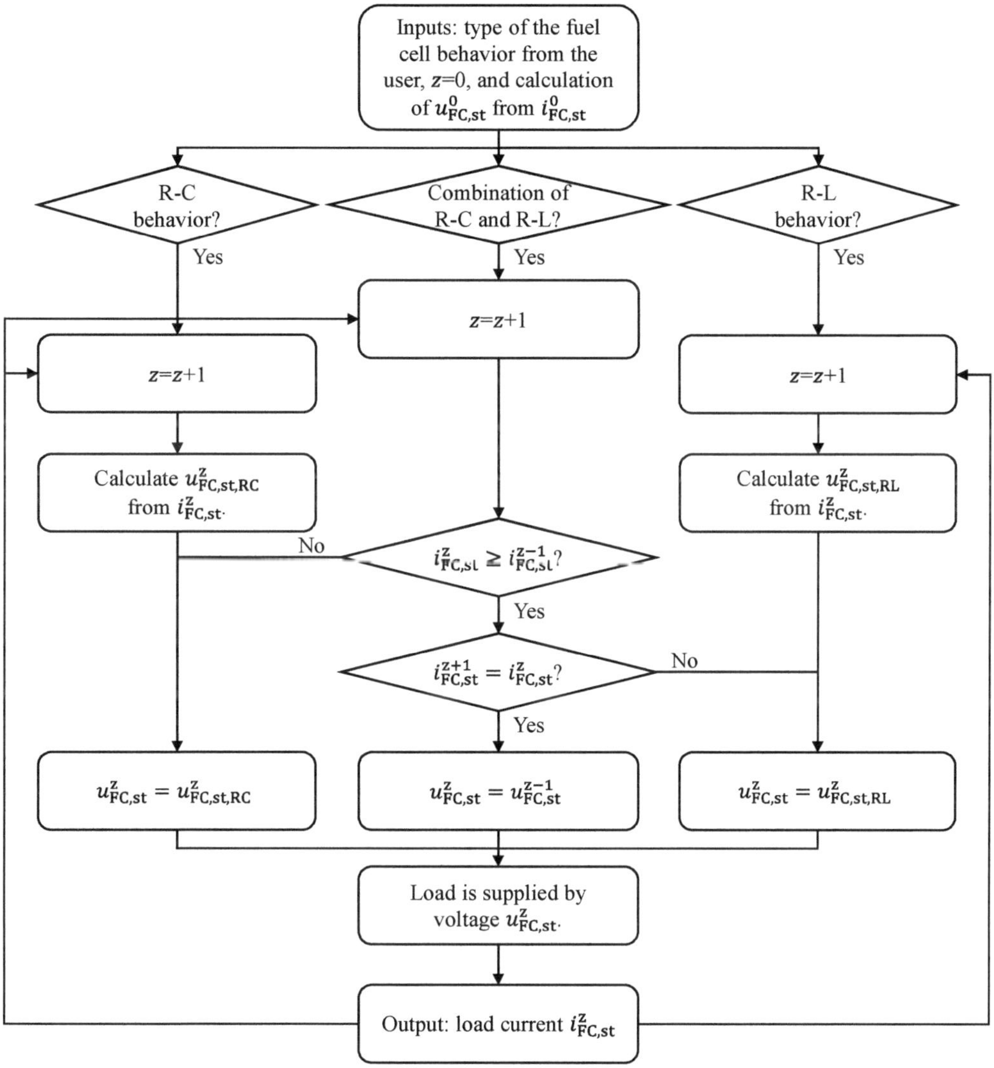

Figure 2.12: Fuel cell transient output voltage selection algorithm based on the user's choice.

2.5 Simulation

2.5.1 Study Case

A fuel cell model based on the above-mentioned approach is developed in MATLAB SIMULINK. Assuming the parameters described in Table 2.4 as experiment data, the model simulates the fuel cell stack considering the stack current variation described in Figure 2.6. These parameters are based on the R-C and R-L models of the fuel cell depicted in Figure 2.8 and Figure 2.10.

Table 2.4: Fuel cell stack assumed experiment data.

Parameter	Symbol	Value
Current Change from 0 A	I	1.52 kA
Minimum Transient Voltage (in the case of R-L behavior)	$U_{\mathrm{tr,min}}$	200 V
Time to Reach Steady-state	t_{ss}	1 s

2.5.2 Results

Figure 2.13 shows the fuel cell simulation results in the case of the R-C and R-L behaviors, while the results for a model with the combination of both behaviors are depicted in Figure 2.14. In the simulation of the fuel cell with combined R-C and R-L behaviors, it is assumed that when the current increases, the fuel cell exhibits R-L behavior, and when the current decreases, it displays R-C characteristics. This model allows users to simulate any fuel cell with any potential behavior.

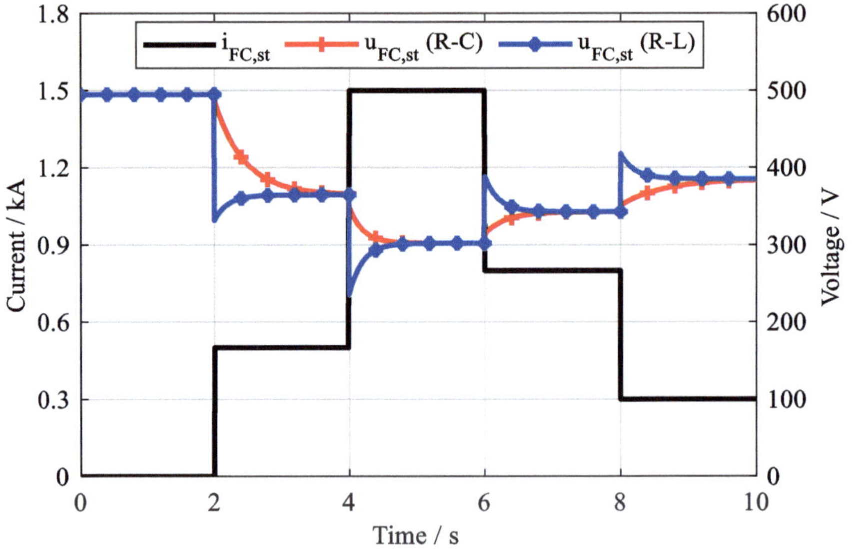

Figure 2.13: Fuel cell stack simulation results considering resistive-capacitive or resistive-inductive behaviors as a response to a current change (calculated with data from Table 2.3 and Table 2.4 and equivalent circuits shown in Figure 2.7 and Figure 2.9).

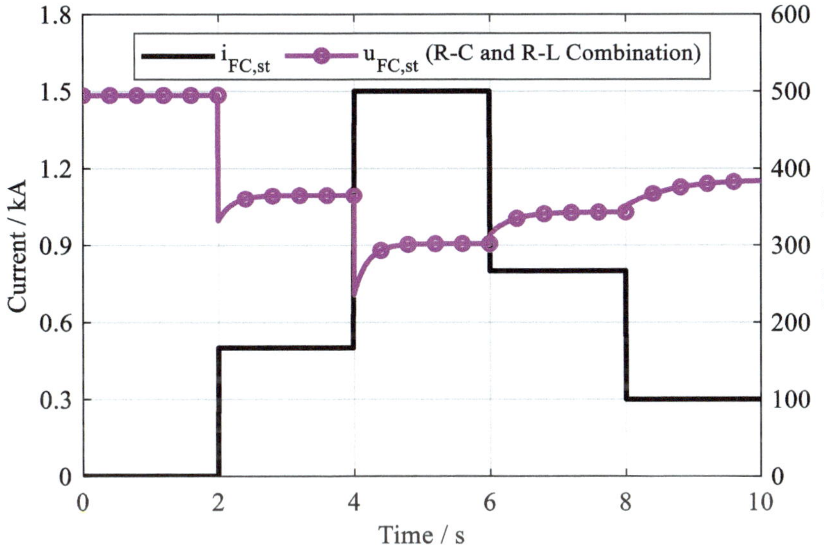

Figure 2.14: Fuel cell stack simulation results considering a combination of resistive-capacitive and resistive-inductive behaviors as a response to a current change (calculated with data from Table 2.3 and Table 2.4 and the algorithm shown in Figure 2.11).

These models are recommended to use either R-C or R-L models in fast current variations like short circuit. Furthermore, a fuel cell system can show different behaviors due to its chemical characteristics, especially when reactants parameters, like pressure or concentration, change. These behaviors need to be examined with experiments. The developed models are the solutions for the fuel cell simulation when a system point of view is intended that is also the objective of this work.

3 Resistive Superconducting Fault Current Limiter

3.1 Background

Limiting fault current and protecting the equipment in an electrical network plays a major role in maintaining stability in the network [81], [82]. Superconducting fault current limiters are promising solutions to limit short circuit current in a significantly fast, reliable, and effective way [81], [83]. These advantages, alongside the significantly low impedance of superconducting fault current limiters during nominal operation, fast fault current limitation, and automatic recovery to the initial state, make them an appealing alternative to conventional current limiting methods [83], [84].

There are various types of fault limiters, such as resistive SFCL, saturated inductive SFCL, and shielded inductive SFCL [85], [86]. Among all types of SFCL, resistive SFCL (RSFCL) is the simplest and most mature solution in comparison to others [87]. Due to the general advantages of FCLs and the benefits of the RSFCLs, with the recent objective of aircraft electrification, there have been interests in using RSFCLs in the powertrain of an electric aircraft [88], [89], [90]. Therefore, this work considers a resistive type fault current limiter to be implemented in the aircraft powertrain.

3.2 High-Temperature Superconducting Tapes

3.2.1 Electric Field, Current Density Characteristic of a Superconductor

In a normal conductor, there is a linear relationship between the electric field and current density, while in a superconductor, in the superconducting state, the electric field is zero. Near the critical current, the electric field begins to change non-linearly with variation in current density [91]. This dependency of electric field E and current density J in a superconductor is called power law. The E-J power law applied to an HTS material is graphically depicted in Figure 3.1 [92].

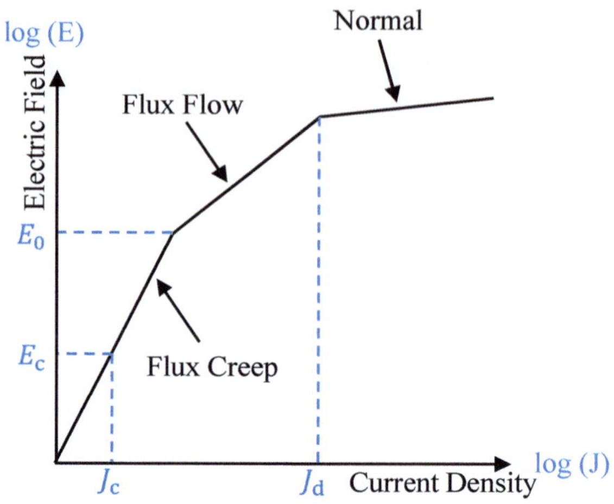

Figure 3.1: Logarithmic electric field-current density curve of the HTS material [92].

Equation (3.1) shows the mathematical description of E-J power law [92], [93].

$$E(J) = E_\mathrm{c} \cdot \left(\frac{J}{J_\mathrm{c}(T)}\right) \cdot \left(\frac{J}{J_\mathrm{c}(T)}\right)^{n-1} \tag{3.1}$$

In this equation, E_c represents the critical electric field. In the case of a superconductor, it is mostly $1\ \mu\mathrm{V/cm}$ [93]. Moreover, n is called the transition index that specifies the steepness of electric field variation depending on current density change. Generally, n is affected by electric field and temperature [93]. Therefore, it is not necessarily a constant parameter. Furthermore, $J_\mathrm{c}(T)$ in equation (3.1) is the critical current density as a function of temperature T. The dependence of critical current with temperature can be calculated in the interval from 65 K to the critical temperature T_c with equation (3.2).

$$J_\mathrm{c}(T) = J_{\mathrm{c},0} \cdot \left(\frac{T_\mathrm{c} - T}{T_\mathrm{c} - T_0}\right) \tag{3.2}$$

In this equation, T_0 is the temperature of cooling fluid. This work considers liquid nitrogen (LN_2) as a coolant at 77 K. Moreover, T_c is the superconductor critical temperature, and $J_{\mathrm{c},0}$ is the initial critical current density at T_0, 77 K.

The specific resistivity of a material ρ is calculated with equation (3.3) based on equation (3.1).

$$\rho = \frac{E}{J} = \left(\frac{E_\mathrm{c}}{J_\mathrm{c}(T)}\right) \cdot \left(\frac{J}{J_\mathrm{c}(T)}\right)^{n-1} \tag{3.3}$$

Using equation (3.3), the resistance can then be calculated. Equation (3.4) shows the basic principle of resistance calculation.

$$R = \rho \cdot \frac{l}{A} \tag{3.4}$$

In this equation, R is the resistance, l represents the conductor length, and A is its cross-section area, which is calculated from conductor depth (thickness) d and width w with equation (3.5).

$$A = d \cdot w \tag{3.5}$$

In Figure 3.1, the transition index n is not constant for any value of the current density J of a superconductor [94]. Therefore, depending on the value of n, E-J behavior of a superconducting material can be divided into three regions [93]. These regions are depicted in Figure 3.1 and explained as follows [91], [93]:

- **Flux Creep**: At this stage, the material is in the superconducting state. The n-value in this region generally lies in the range of 10-50. For example, for YBCO (Yttrium Barium Copper Oxide) tapes, it is between 20 and 30 [95]. In a well-designed model, the current density of a superconductor is smaller than the critical current density. Thus, based on equation (3.3), the resistance of a superconductor is very small (almost zero) at this stage.
- **Flux Flow**: This stage represents an extension of the flux creep region. Typically, the superconductor starts to enter this region upon a rise in its current and temperature. This occurs when the electric field across the superconductor exceeds an E_0 which is typically in the range of 0.1 to 10 mV/cm. The value of n in this region is much less than the flux creep region. It is usually between 2 and 4 [96]. Considering equation (3.3), the resistance of the superconductor is still relatively small, but it is higher than the flux-creep stage.
- **Normal**: By rising the temperature of the superconductor to more than the critical temperature, the superconductor is no longer in the superconducting state. A similar effect occurs if the current density surpasses the so-called "de-pairing current density (J_d)" [97]. In this state, the superconductor behaves like a normal conductor (resistor), with a transition index of 1. As a result, the superconductor's specific resistivity can be calculated as a linear function of temperature T, as shown in equation (3.6). In this equation, m and ρ_0 are the material-specific data to calculate their resistivity.

$$\rho = m \cdot T + \rho_0 \tag{3.6}$$

3.2.2 Properties of the Tapes

In this work only REBCO high-temperature superconductors are considered with critical temperatures above 90 K and critical magnetic fields of more than 100 T. The abbreviation RE stands for rare earth, and depending on the manufacturer, several materials like Yttrium or Gadolinium are possible candidates. In this work, the superconducting tape is represented with three layers of material: one layer for the superconductor and the other two layers are Hastelloy as a substrate and silver as a stabilizer. In reality, there are more layers, such as thin buffer layers, but due to their small dimensions and very high resistance, they can be neglected in this simulation. The general structure of a REBCO tape is depicted in Figure 3.2.

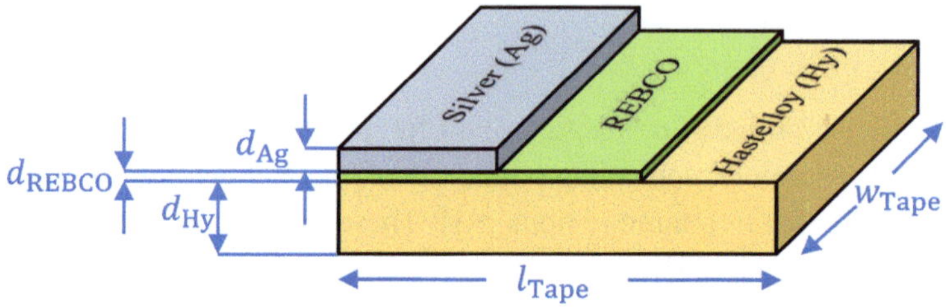

Figure 3.2: General structure of a superconducting REBCO tape considered for the SRSFCL design with the properties described in Table 3.1.

Table 3.1 shows the typical dimensions related to each layer of the chosen superconducting tape alongside its critical current and temperature. The symbols regarding tape dimensions for each layer are illustrated in Figure 3.2.

Table 3.1: Properties of the superconducting tape considered for the RSFCL design with the structure shown in Figure 3.2.

Parameter	Symbol	Value
Tape Width	w_{Tape}	1.2 cm
Tape (Length)	l_{Tape}	8.5 m
Silver Thickness	d_{Ag}	4 µm
REBCO Thickness	d_{REBCO}	1 µm
Hastelloy Thickness	d_{Hy}	100 µm
Tape Total Thickness	d_{Tape}	105 µm
Tape Cross-section Area	A_{Tape}	1.26 mm^2
Tape (REBCO) Critical Current	I_c	550 A @ 77 K
Tape (REBCO) Critical Temperature	T_c	92 K

3.3 Design and Properties

The critical temperature of REBCO is around 92 K [92], thus the tapes made up of this superconductor must be cooled to lower than this temperature. To provide such a temperature, liquid nitrogen (LN2) operating at 77 K is used. A SFCL is constructed using multiple tapes arranged in parallel. In nominal conditions, since the resistance of REBCO is much smaller than other layers, it can be expected with a high accuracy that nearly all the current flows through this layer. The tapes must be capable of handling the maximum nominal current that passes through them. As described in Chapter 1, Table 1.1, the maximum nominal DC current is 1.52 kA. Considering the tapes critical current listed in Table 3.1, which is 550 A, at least three tapes in parallel are required.

3.4 Electrical Characteristic

According to the previous sub-chapter, each layer of the HTS tape has its own material dependent resistance. Therefore, a superconducting tape can be modeled as a resistor. The electrical model of a superconducting tape is shown in Figure 3.3. The resistance of a superconducting tape, R_{Tape}, would be calculated based on the parallel resistors law. This calculation is found in equation (3.7).

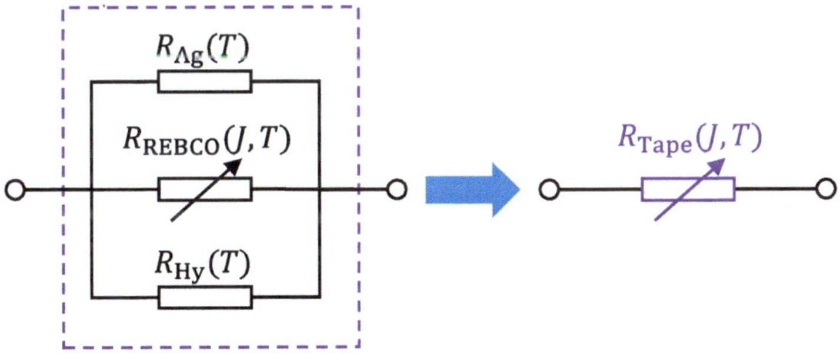

Figure 3.3: Electrical model of an HTS tape, considering the resistances of the layers.

$$\frac{1}{R_{\mathrm{Tape}}} = \frac{1}{R_{\mathrm{Ag}}} + \frac{1}{R_{\mathrm{REBCO}}} + \frac{1}{R_{\mathrm{Hy}}} \tag{3.7}$$

In this equation, R_{Ag} and R_{Hy} are the silver and Hastelloy layers resistances. These materials are normal resistors at 77 K and above, hence, their resistances vary with only temperature T based on equation (3.6). However, as mentioned previously, REBCO layer resistance R_{REBCO} changes non-linearly with temperature T and the density of the current flowing through it, J. The specific data to calculate the resistance of these materials can be found in the appendices Chapter 9, section A.1, Table 9.1.

As mentioned in the previous sub-chapter, an RSFCL comprises several parallel HTS tapes. Considering N_{Tape} number of tapes to construct the RSFCL with identical properties and the same temperature and current shared between them, the total resistance of the HTS tapes $R_{\mathrm{Tape,tot}}$ is calculated with equation (3.8).

$$R_{\mathrm{Tape,tot}} = \frac{R_{\mathrm{Tape}}}{N_{\mathrm{Tape}}} \tag{3.8}$$

3.5 Thermal Characteristic

In addition to the electrical behavior, an HTS tape exhibits a thermal characteristic which is influenced by fluctuations in its temperature. As shown in equation (3.2), the temperature variation causes a change in the tape critical current, ultimately leading to its resistance alteration. The main source of a change in the RSFCL tapes temperature is the power loss within the tapes. When a continuous current with an amplitude I_{Tape} flowing through a tape with a resistance R_{Tape}, the power loss in the tape, P_{Tape}, is determined with equation (3.9). This power loss is directly responsible for heating the tapes and affecting their thermal behavior.

$$P_{\text{Tape}} = R_{\text{Tape}} \cdot I_{\text{Tape}}^2 \tag{3.9}$$

Depending on the heat exchange between tapes and surrounding areas like the coolant liquid or the energy exchange between the layers, the characteristic modeling of SFCL is significantly different. The tape temperature T_{Tape} varies over time t by the power loss in the tape P_{Tape} and the convected power to the cooling media, P_{c}. This is shown in equation (3.10), where the parameter C_{Tape} represents the tape heat capacity. The calculation of this parameter alongside the specific heat capacities of all the layers in the HTS tape are explained in the appendices Chapter 9, section A.1.

$$C_{\text{Tape}} \cdot \frac{\partial T_{\text{Tape}}}{\partial t} = P_{\text{Tape}} - P_{\text{c}} \tag{3.10}$$

Based on equation (3.10) and using Euler method [98], the temperature change in the tapes ΔT_{Tape} is calculated with equation (3.11), where Δt represent the simulation time-step.

$$\Delta T_{\text{Tape}} = \Delta t \cdot \left[\frac{P_{\text{Tape}} - P_{\text{c}}}{C_{\text{Tape}}}\right] \tag{3.11}$$

This equation shows that in nominal conditions with a current passing by superconducting tape less than the critical current I_{c}, its temperature slightly increases. In the opposite, in case of a sudden rise in this current (for example, due to a fault), the temperature increases significantly, possibly to more than the critical temperature T_{c}. As explained before, the specific resistivity ρ is dependent on the temperature. Therefore, the resistance of the tape itself changes with the temperature variation.

Moreover, depending on the thermal behavior considerations, two cases can be analyzed for the cooling power P_{c}: adiabatic environment, and non-adiabatic environment. These cases are explained below.

3.5.1 Adiabatic Environment

Tapes are considered in an adiabatic environment if there is no heat exchange between them and the coolant liquid. Hence, in this model, the convected heat rate is considered to be zero ($P_{\mathrm{c,adiabatic}}$=0 W). This case is interesting to analyze as it represents a worst-case scenario where coolant loses its capability to cool down the tapes.

3.5.2 Non-adiabatic Environment

On the other hand, HTS tapes are considered in a non-adiabatic environment if they have a heat exchange with the cooling media, LN_2. As already mentioned before, the temperature throughout the whole tape is considered the same as a single temperature T_{Tape}. Therefore, no heat is exchanged between layers of the tape, and only a heat transfer between the HTS tape and the cooling media occurs at the surfaces of the tape. The total convected power P_{c} between the HTS tape and the cooling media is graphically shown in Figure 3.4. This power includes the convection from both surfaces of the HTS tape and is determined using equation (3.12).

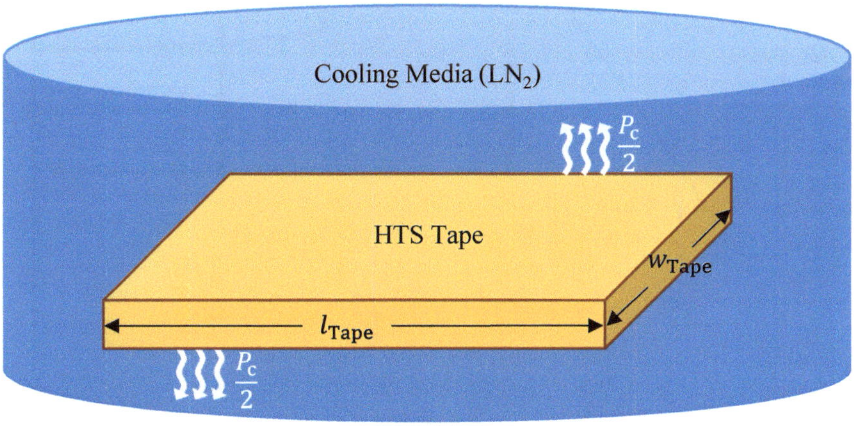

Figure 3.4: Graphical illustration of the heat convection between an HTS tape and the cooling media in the non-Adiabatic model.

$$P_{\mathrm{c,non-Adiabatic}} = 2h_{\mathrm{c}} \cdot w_{\mathrm{Tape}} \cdot l_{\mathrm{Tape}} \cdot \left[T_{\mathrm{Tape}} - T_{\mathrm{LN_2}}\right] \tag{3.12}$$

The parameter h_{c} (unit: W/(m^2.K)) in this equation is the coolant convective heat transfer coefficient. This parameter specifies the amount of heat exchanged between one m^2 area of the HTS tape surfaces and liquid nitrogen at the room pressure in the span of time. Since each coolant shows a different thermal behavior, their parameter h_{c} is also different. In the case of an RSFCL, tapes are cooled in a bath of coolant (pool cooling) and a constant temperature for the coolant is assumed. To compare coolants on this perspective, the calculation of the LN_2 at 77 K and LH_2 (liquid Hydrogen) at 20 K convective heat transfer coefficients, $h_{\mathrm{c,LN_2}}$ and

h_{c,LH_2}, are shown in equations (3.13) and (3.14) [92], [99]. In these equations, ΔT is the difference between the coolant temperature and the temperature at the surface of the tapes. The α_i parameters for the calculation of the h_{c,LN_2} are given in the appendices Chapter 9, section A.3. Furthermore, Figure 3.5 shows the h_c profiles for these two coolants.

$$h_{c,LN_2} = \begin{cases} 2170 & ; \Delta T \le 3.2 \\ (\alpha_0 + \alpha_1 \Delta T + \alpha_2 \Delta T^2 + \alpha_3 \Delta T^3 + \alpha_4 \Delta T^4 + \alpha_5 \Delta T^5)/\Delta T & ; 3.2 < \Delta T \le 28 \\ (3970 + 98\Delta T)/\Delta T & ; \Delta T > 28 \end{cases} \tag{3.13}$$

$$h_{c,LH_2} = \begin{cases} 100\Delta T^{5.3} & ; \Delta T \le 3 \\ \dfrac{10^5}{\Delta T} & ; 3 < \Delta T \le 100 \\ 1000 & ; \Delta T > 100 \end{cases} \tag{3.14}$$

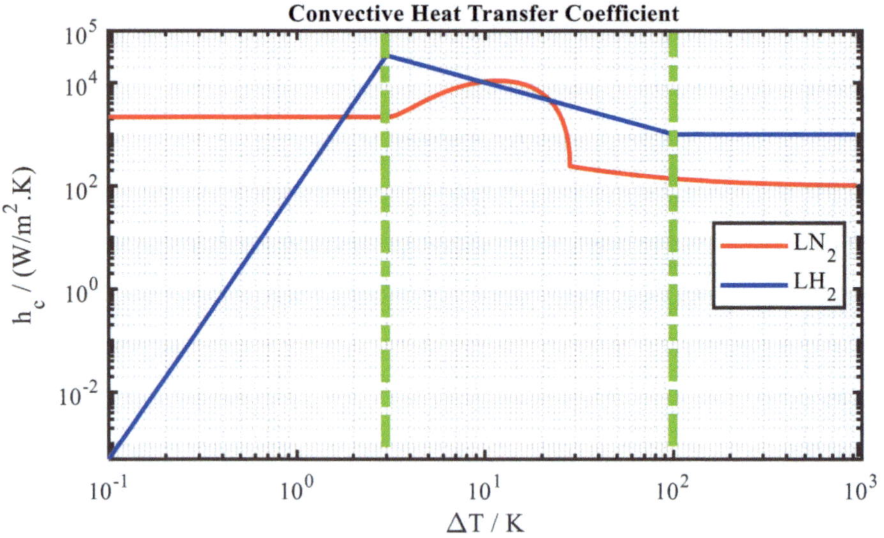

Figure 3.5: Profiles of the coolants convective heat coefficients, h_c: LN_2 at 77 K, and LH_2 at 20 K [92], [99].

This figure illustrates that during a fault, the increase in the tapes temperature, and consequently the rise in the temperature difference between the tapes and the coolant, affects how the cooling fluids absorb heat. When the temperature increases relatively low, liquid nitrogen (LN_2) performs better in absorbing heat. However, when the temperature rise is more significant, liquid hydrogen (LH_2) demonstrates superior heat absorption, which likely allows it to restore the tapes to their initial superconducting state more rapidly.

3.6 Non-linear, Adiabatic, and Non-adiabatic, Lumped-parameter Model

The RSFCL modeling must take both characteristics of a fault current limiter into account. As it was already pointed out in previous sub-chapters, an R-SFCL has multiple factors which strongly affect each other. In this work, the modeling of a fault limiter using the lumped-parameter method is proposed. This method considers all the parameters explained before.

Figure 3.6 shows the algorithm flowchart of RSFCL simulation with the non-linear, lumped parameter model. In this scheme, the number of each step can be found at the bottom left corner, and each time-step is shown with z.

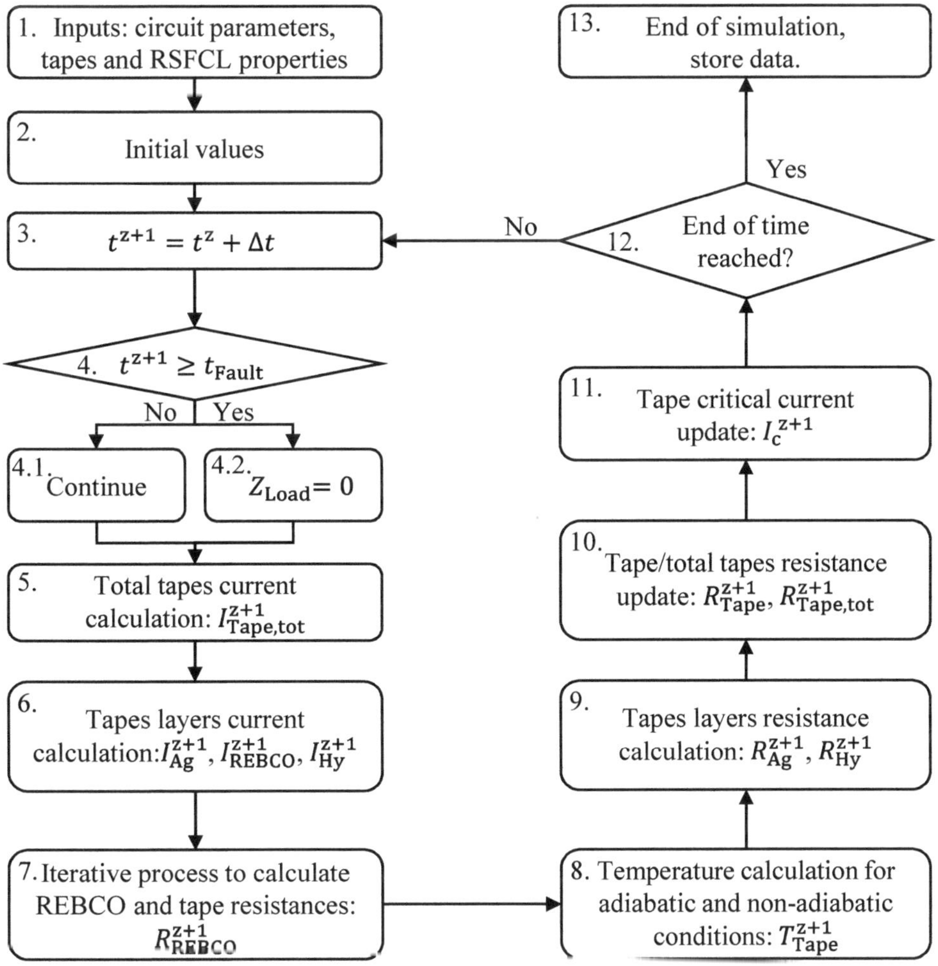

Figure 3.6: Algorithm flowchart of the RSFCL modeling [92].

This lumped parameter model allows adiabatic and non-adiabatic environments to be further distinguished. Each step of the RSFCL modeling is explained below in detail.

1. At this step, the network parameters, the tapes, and RSFCL properties are taken from the user.

2. At this step, the initial values of different parameters, including tapes temperature and resistance and their sub-layers are calculated based on the approaches explained in previous sub-chapters. These values specify the status of the network component at the beginning of the simulation.

3. The simulation in a time-loop starts from this step, where parameters like simulation time and time-step interval Δt are taken. The parameter $t^{\mathrm{z+1}}$ represents current (new) time-step while t^{z} indicates the previous time-step.

4. At this step, the existence of a fault is checked. In the case of a fault at a time t_{Fault}, the load impedance Z_{Load} is short-circuited (Z_{Load}=0). If no fault is seen, no change in the network is applied. It is worth mentioning that this step is only applicable when the RSFCL is simulated in the standalone mode to analyze its behavior. When it is implemented with other components, this step is skipped, and RSFCL model reacts automatically to a fault.

5. Next, the total current passing through all tapes at current time-step, $I_{\mathrm{Tape,tot}}^{\mathrm{z+1}}$, must be calculated. Since RSFCL is a non-linear component, this current cannot be found analytically. Thus, a numerical method to estimate its value is needed. As an example, the 5th order Runge-Kutta-Fehlberg method can be used [100].

6. By calculation of the total current and assuming identical properties for all N_{Tape} tapes, the current is distributed evenly among them. The current in each sub-layer can then be determined. For instance, the REBCO current at current time-step, $I_{\mathrm{REBCO}}^{\mathrm{z+1}}$, is derived from its resistance at previous time-step, $R_{\mathrm{REBCO}}^{\mathrm{z}}$ with equation (3.15). In this equation, $R_{\mathrm{Tape}}^{\mathrm{z}}$ represents the tape resistance at previous time-step, which is calculated using equation (3.7).

$$I_{\mathrm{REBCO}}^{\mathrm{z+1}} = \frac{R_{\mathrm{Tape}}^{\mathrm{z}}}{R_{\mathrm{REBCO}}^{\mathrm{z}}} \cdot \frac{I_{\mathrm{Tape,tot}}^{\mathrm{z+1}}}{N_{\mathrm{Tape}}} \tag{3.15}$$

7. As explained in previous sub-chapters, a superconductor has a non-linear resistance affected by current and temperature. This makes it challenging to accurately determine its resistance, particularly when the current exceeds the critical current I_{c} but the temperature remains below the critical temperature T_{c}. Therefore, a reliable algorithm is required to calculate the precise REBCO resistance, $R_{\mathrm{REBCO}}^{\mathrm{z+1}}$. The algorithm for this calculation can be found in [92].

8. In the next step, the tapes temperature $T_{\mathrm{Tape}}^{\mathrm{z+1}}$ is calculated for both adiabatic and non-adiabatic conditions. The approach is explained in previous section.

9. With new tapes temperature, the resistances of other sub-layers in the tapes must be updated as well. Hence, new resistance of silver and Hastelloy layers, $R_{\mathrm{Ag}}^{\mathrm{z+1}}$ and $R_{\mathrm{Hy}}^{\mathrm{z+1}}$, are calculated with equation (3.6).

10. Since the new resistances of all sub-layers are updated, the tapes resistance and, consequently, total resistance of all tapes are also updated based on equations (3.7) and (3.8).

11. Furthermore, due to the temperature change, the tapes critical current must be updated using equation (3.2).

12. Once all parameters for the current time step have been calculated, the simulation should proceed to the next step if the desired simulation time has not yet been reached. If the simulation time has been completed, the process concludes.

13. In the final step, data such as the tapes temperature and resistance are stored for future reference and analysis.

3.7 Simulation

To simulate the models described in previous sections, MATLAB programming is used. Moreover, a MATLAB SIMULINK block has been developed, offering flexibility to adjust parameters and select between the adiabatic and non-adiabatic models [101]. This section considers a basic study case to analyze the RSFCL impact on a network in nominal conditions and short circuit events.

3.7.1 Study Case

This study focuses on simulating the RSFCL within a bipolar DC network powered by a constant DC voltage source. The schematic of this network is illustrated in Figure 3.7, and the detailed specifications of the network components are provided in Table 3.2. As it is a bipolar configuration, it is assumed that both poles have identical line impedance, with the total resistance and inductance of each pole represented by R_{Line} and L_{Line}, respectively. A pole-to-pole short circuit scenario is considered to occur at the load location, where the most severe case with 0 Ω fault resistance is assumed. The parameters related to this short circuit event are also outlined in Table 3.2.

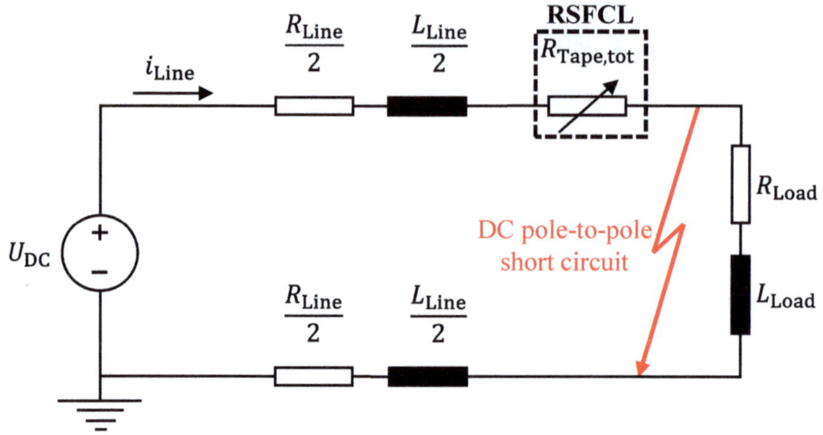

Figure 3.7: Study case network schematic to simulate RSFCL behavior in nominal and short circuit conditions with the components described in Table 3.2.

Table 3.2: Specifications of the elements in the study case network, used to simulate the RSFCL behavior, illustrated in Figure 3.7.

Parameter	Symbol	Value
DC Voltage Source Amplitude	U_{DC}	300 V
Line Resistance	R_{Line}	0.02 Ω
Line Inductance	L_{Line}	40 μH
Load Resistance	R_{Load}	0.18 Ω
Load Inductance	L_{Load}	0.38 mH
Nominal Line DC Current	$I_{\text{Line,nom}}$	1.5 kA
Fault (Short Circuit) Resistance	R_{Fault}	0 Ω
Fault Moment	t_{Fault}	0.1 s
Fault Duration	$t_{\text{d,Fault}}$	100 ms

3.7.2 Results

The line current i_{Line} resulted from the study case simulation is shown in Figure 3.8. This figure illustrates that in the absence of an RSFCL, the fault current can surge up to 15 kA. However, with the RSFCL integrated into the network, the fault current is significantly reduced and limited to around 6 kA, demonstrating the RSFCL effectiveness in mitigating fault conditions.

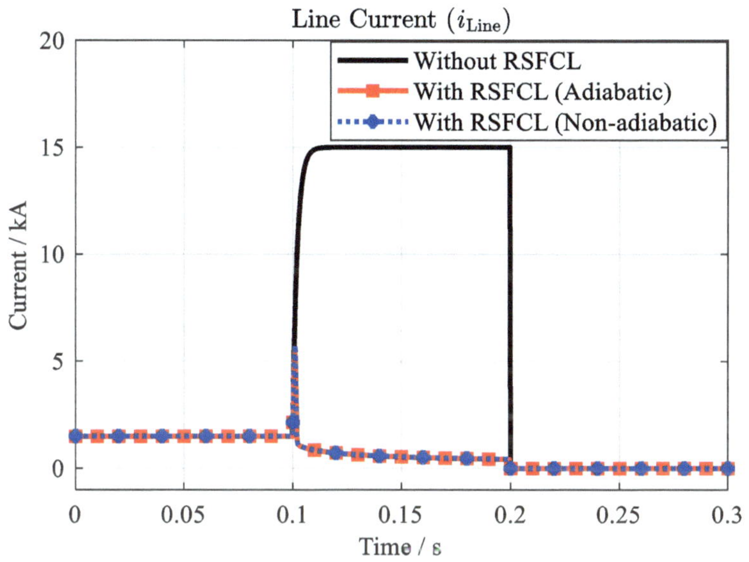

Figure 3.8: Line current resulted from the study case network simulation with the schematic depicted in Figure 3.7 in the conditions with and without an RSFCL.

Another important observation is the negligible difference between the RSFCL performance in the adiabatic and non-adiabatic conditions during fault. It can be concluded that during the fault, cooling of the HTS tapes is not capable of reducing the tapes temperature. It seems that during a fault, the effect of the power loss in the tapes is relatively higher than the convected heat with liquid nitrogen.

Moreover, the tapes temperature T_{Tape} and their total resistance in the RSFCL $R_{\text{Tape,tot}}$ are depicted in Figure 3.9. This figure highlights the RSFCL impact on limiting the fault current while their temperature remains at 77 K in nominal conditions. After the fault, the tapes quench as their temperature rapidly exceeds the critical temperature T_c (92 K) within about 1 ms, due to the power loss in the tapes. This transition is also reflected in the total tape resistance, where they exhibit almost no resistance under nominal conditions but experience a sharp rise in resistance after the fault, effectively limiting the fault current.

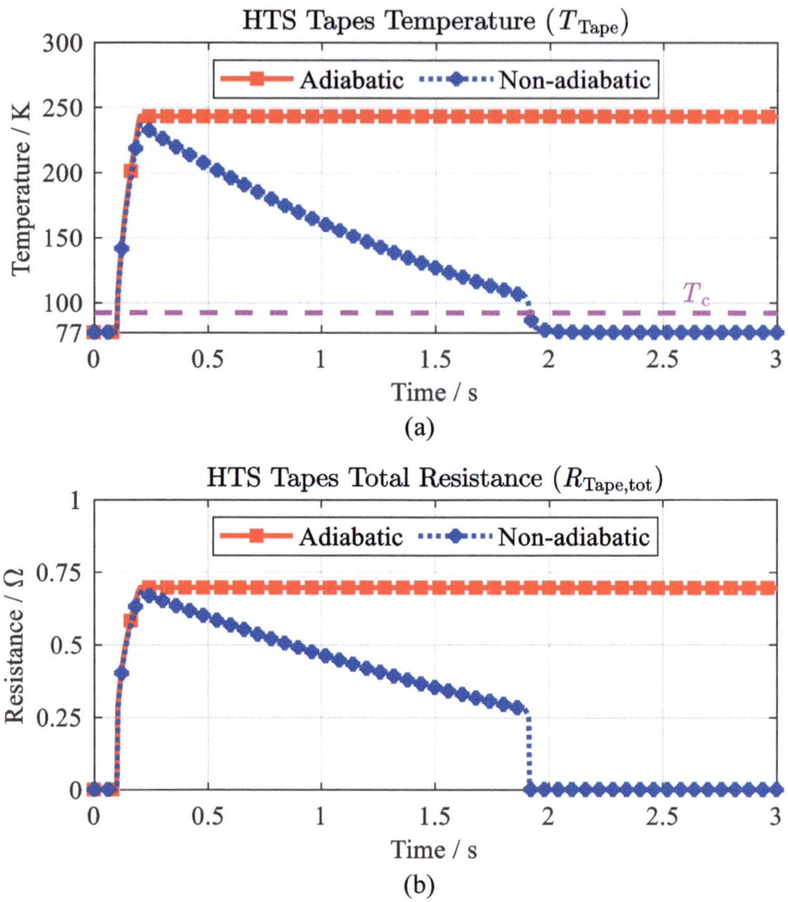

Figure 3.9: Results of the study case network simulation with the schematic depicted in Figure 3.7 in the adiabatic and non-adiabatic conditions: (a) HTS tapes temperature, (b) HTS tapes resistance.

Comparing the results between adiabatic and non-adiabatic conditions reveals that the temperature rise occurs faster than the heat can be absorbed by liquid nitrogen (LN2), showing little difference between the two cases during the fault. However, in the non-adiabatic scenario, it takes approximately 2 seconds for the coolant to absorb all the generated heat and restore the REBCO tapes to their superconducting state.

Another notable observation is regarding the currents of each sub-layer in the HTS tapes, as shown in Figure 3.10. Under nominal conditions, the REBCO layer conducts the entire current due to its extremely low resistance. However, after a fault occurs and the resistance of the REBCO layer rises, causing the tapes to transition to a normal conductive state, the line current shifts to the silver layer, which has a much lower resistance compared to the other layers. The Hastelloy layer carries a minimal current, as its resistance is higher than that of the silver layer.

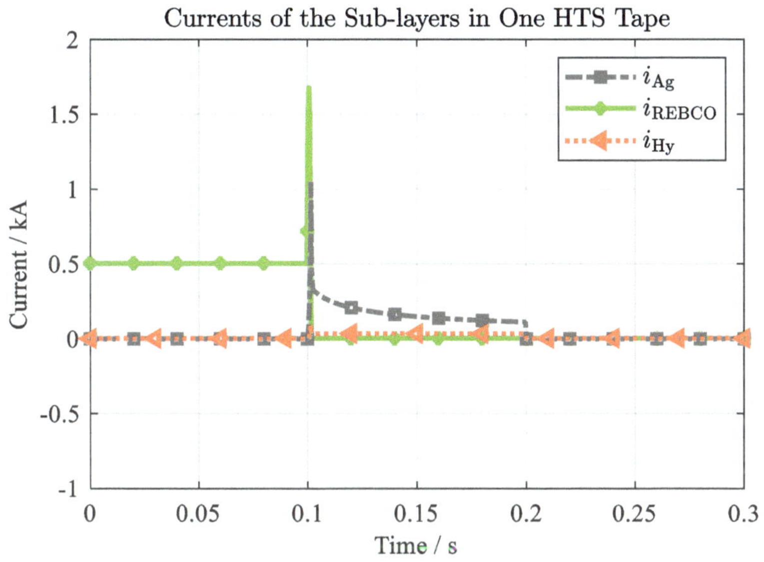

Figure 3.10: Currents of each sub-layer in the HTS tapes in nominal conditions and during fault resulted from the study case network simulation with the schematic depicted in Figure 3.7, with i_{Ag}, i_{REBCO}, and i_{Hy} being the silver, REBCO superconductor, and Hastelloy layers currents, respectively.

The developed RSFCL models are highly adaptable, allowing customization based on specific requirements and parameters. In this work, only REBCO superconducting tapes were discussed. Still, the models can be used with the same principles for other superconductors, like Magnesium diboride (MgB_2) and Niobium Titanium (NbTi) superconducting wires. These models can be applied to both DC and AC networks, providing highly accurate results. Additionally, the models are designed for efficiency; running the study case simulation with an RSFCL model in MATLAB and SIMULINK on a computer with a 6-core 2.3 MHz CPU and 16 GB RAM is nearly instantaneous. This demonstrates their capability for fast performance without sacrificing accuracy.

3.7.3 Models Comparison

The models developed to simulate RSFCL behavior in adiabatic and non-adiabatic conditions can be used in different conditions. On the one hand, the adiabatic model refers to conditions where the coolant has no effect on the tapes, such as a worst-case scenario of cooling fluid loss. On the other hand, the non-adiabatic model is developed to simulate the RSFCL in a standard scenario, where coolant can absorb additional heat generated in the tapes by a short circuit. Simulation results indicated significantly similar results in both cases during a short circuit, meaning that the generated heat in the tapes occurs relatively faster than the coolant ability to dissipate it. However, to figure out the time after the fault when all the generated heat is absorbed by coolant is only a capability of the non-adiabatic model. This time shows when HTS tapes return to the superconducting state and nominal operation. Therefore, its estimation via the non-adiabatic model is essential as it can be used to control and protect the other components.

4 Superconducting Cable

4.1 Background

Superconducting cables have been at the center of attention in recent decades [102], [103], [104]. These cables offer promising advantages such as smaller size, higher current density, lower loss, and potential fault current limiting capability compared to conventional cables [103], [104], [105]. These advantages have been a solid motivation to move forward to commercialize superconducting cables and use them in various power systems [102], [103], [106].

There have been several projects for the design, manufacturing, and testing of superconducting cables [107], [108], [109], [110], [111], [112], including the AmpaCity and SuperLink projects in Germany [113], [114], [115]. Due to the attractiveness and potential advantages of the superconducting cables, they have a variety of potential applications in submarine, railway, and electric aircraft power systems [104], [116], [117], [118], [119]. Especially in aviation-related applications in which weight and space are crucial criteria, superconducting cables seem to be an appealing solution for their lightweight and compact properties. In this regard, the research in [120] shows that, for an electric powertrain, a superconducting cable with a weight density as low as 0.35 kg/kA/m (coolant included) can be designed. Thus, in a high-power electric aircraft, superconducting cables present a solution for carrying high currents while maintaining low voltage levels.

This work explains the electrothermal modeling of superconducting DC and AC cables. The properties and structure of each cable are provided. All the developed models are independent of the design and type of the cable, and adaptable with requirements.

4.2 Design and Properties

4.2.1 High-temperature Superconducting Tapes Properties

Similar to the RSFCL, an HTS tape made up of REBCO tapes is considered in the DC and AC cable design. For a precise analysis, the 4-mm tapes with a critical current I_c of 210 A and 100 A are chosen respectively for the DC and AC cables [121]. These superconducting tapes are represented with four layers of material: one layer for the REBCO superconductor, the Hastelloy (Hy) layer as a substrate, and the silver (Ag) and copper (Cu) layers as stabilizer. Figure 4.1 illustrates the structure of the HTS tapes, and its properties are described in Table 4.1.

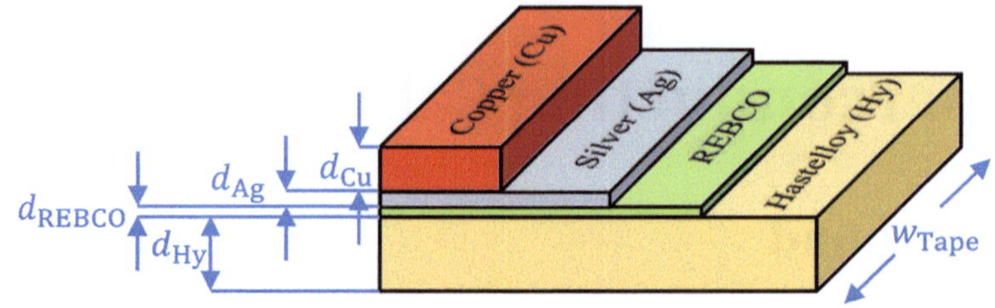

Figure 4.1: Physical structure of the HTS tape considered for the DC and AC cables design with the properties described in Table 4.1 [121], [122].

Table 4.1: Properties of the HTS tapes considered for the DC and AC cables design with the structure shown in Figure 4.1 [121], [122].

Parameter	Symbol	Value	
		DC Cable	AC Cable
Tape Width	w_{Tape}	4 mm	4 mm
Copper Thickness	d_{Cu}	40 µm	9 µm
Silver Thickness	d_{Ag}	3 µm	2.4 µm
REBCO Thickness	d_{REBCO}	2 µm	1.6 µm
Hastelloy Thickness	d_{Hy}	75 µm	50 µm
Tape Total Thickness	d_{Tape}	120 µm	63 µm
Tape Cross-section Area	A_{Tape}	0.48 mm^2	0.25 mm^2
Tape (REBCO) Critical Current	I_{c}	210 A @ 77 K	100 A @ 77 K
Tape (REBCO) Critical Temperature	T_{c}	92 K	92 K

4.2.2 Cable Geometry

This work considers a bipolar coaxial DC cable with the structure depicted in Figure 4.2 and the geometry described in Table 4.2. In this cable, the HTS tapes are cooled by LN_2 at 77 K, which is pumped from one side and flows in one duct on the outside, while both poles are placed in one cryostat. The inner pole consists of 8 HTS tapes arranged in parallel, with a core made up of copper also as former. The outer pole contains 12 HTS tapes in parallel and includes a shield copper. In these two poles, the HTS tapes are electrically in parallel with the copper layers. It is important to note that the difference in the number of HTS tapes between the poles is due to the larger outer diameter of the outer HTS tapes. To fully occupy the available space in the cable, more tapes are required in the outer pole. Additionally, a Polypropylene Laminated Paper (PPLP) is placed between the poles for electrical insulation.

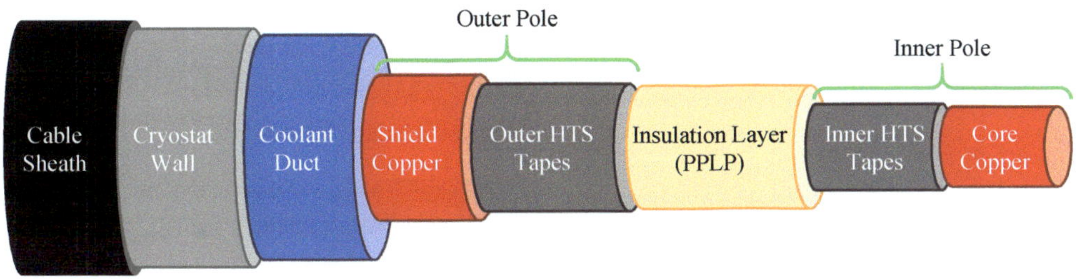

Figure 4.2: Structure of the coaxial bipolar HTS DC cable composed of HTS tapes, as described in Figure 4.1 and Table 4.1, with the properties specified in Table 4.2.

Table 4.2: Properties of the coaxial bipolar HTS DC cable composed of HTS tapes, as described in Figure 4.1 and Table 4.1, with the structure illustrated in Figure 4.2.

Parameter	Symbol	Value
Core Copper Diameter	D_{Core}	10 mm
Inner HTS Tapes Outer Diameter	$D_{\mathrm{Tape,in}}$	10 mm
PPLP Outer Diameter	D_{PPLP}	15 mm
Outer HTS Tapes Outer Diameter	$D_{\mathrm{Tape,out}}$	15 mm
Shield Copper Outer Diameter	D_{Sh}	18 mm
Cooling Fluid (LN_2) Outer Diameter	D_{Fld}	28 mm
Cable Diameter	D_{Cable}	30 mm
Cable Length	l_{Cable}	50 m
Number of the Inner HTS Tapes	$N_{\mathrm{Tape,in}}$	8
Inner HTS Tapes Total Critical Current	$I_{\mathrm{c,Tape,in}}$	1.68 kA
Number of the Outer HTS Tapes	$N_{\mathrm{Tape,out}}$	12
Inner HTS Tapes Total Critical Current	$I_{\mathrm{c,Tape,out}}$	2.52 kA
Cooling Fluid Mass Flow Rate	$\dot{m}_{\mathrm{Fld}}$	0.4 kg/s

To deliver the electrical energy on the AC side from DC/AC inverter to the motor, a three-phase AC cable with the structure depicted in Figure 4.3 and the geometry described in Table 4.3 is integrated. In this cable, the HTS tapes are cooled with LN_2 at 77 K, pumped from one side and flows in cryostat where all phases are placed. Each phase comprises eight parallel layers of HTS tapes, each containing three tapes arranged on an aluminum core, which also serves as a former. The cable includes a copper shield which is not as thick as the shield copper used in the DC cable. A Kapton dielectric layer is also placed between the poles for electrical insulation.

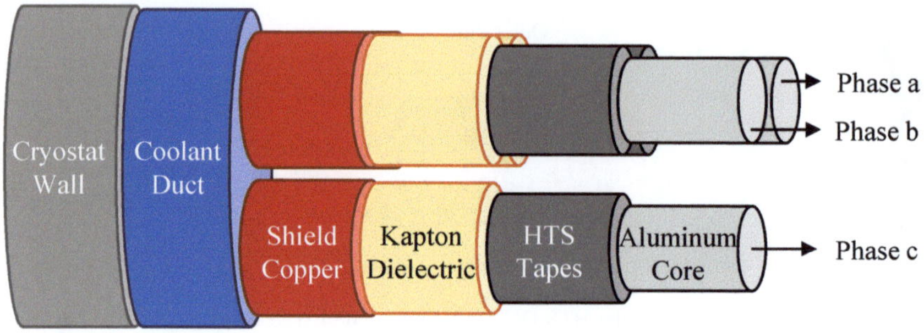

Figure 4.3: Structure of the coaxial three-phase HTS AC cable composed of HTS tapes, as described in Figure 4.1 and Table 4.1, with the properties specified in Table 4.3.

Table 4.3: Properties of the coaxial three-phase HTS AC cable composed of HTS tapes, as described in Figure 4.1 and Table 4.1, with the structure illustrated in Figure 4.3.

Parameter	Symbol	Value
Aluminum Core Diameter	D_{Core}	5.5 mm
HTS Tapes Outer Diameter	D_{Tape}	6.5 mm
Kapton Outer Diameter	D_{Ka}	7.4 mm
Shield Copper Outer Diameter	D_{Sh}	7.5 mm
Cooling Fluid (LN_2) Outer Diameter	D_{Fld}	39 mm
Cryostat Outer Diameter	D_{Cryo}	66 mm
Cable Length	l_{Cable}	2 m
Number of the HTS Tapes Layer per Phase	$N_{\mathrm{Tape,layer,ph}}$	8
Number of the HTS Tapes per Layer per Phase	$N_{\mathrm{Tape/layer,ph}}$	3
Total Number of the HTS Tapes per Phase	$N_{\mathrm{Tape,ph}}$	24
HTS Tapes Total Critical Current per Phase	$I_{\mathrm{c,ph}}$	2.4 kA
Cooling Fluid (LN_2) Temperature	$T_{\mathrm{LN_2}}$	77 K

4.2.3 Cable Inductance

In addition to the cable resistance, calculation of the superconducting cable inductance is necessary. The cable inductance is strongly dependent on the structure of the cable. For the coaxial DC cable described in Figure 4.2, each pole of the bipolar cable has its own total effective inductance L_{e} which includes as well the mutual inductance with the other pole. If $L_{\mathrm{e,in}}$ and $L_{\mathrm{e,out}}$ are respectively the total effective inductances of inner and outer poles, the total cable inductance L_{Cable} can be calculated with equation (4.1).

$$L_{\mathrm{Cable}} = L_{\mathrm{e,in}} + L_{\mathrm{e,out}} \tag{4.1}$$

Knowing this equation, the total cable inductance with the parameters given in Table 4.2 is calculated via equation (4.2) [123]:

$$L_{\text{Cable}} = \frac{\mu_0 \cdot l_{\text{Cable}}}{2\pi} \cdot \left[\ln\left(\frac{D_{\text{Sh}}}{D_{\text{Tape,in}}}\right) + \frac{2\left(\frac{D_{\text{PPLP}}}{d_{\text{Sh}}}\right)^2}{1-\left(\frac{D_{\text{PPLP}}}{D_{\text{Sh}}}\right)^2} \cdot \ln\left(\frac{D_{\text{Sh}}}{D_{\text{PPLP}}}\right) - \frac{3}{4} + \ln(\zeta) \right] \quad (4.2)$$

In this equation, μ_0 is the free space permeability and it is equal to $4\pi \times 10^{-7}$ $[\text{T.m/A}]$. Moreover, the parameter ζ is dependent on the thickness of the outer pole, and ratio $D_{\text{PPLP}}/D_{\text{Sh}}$. In the given cable, parameter $\ln(\zeta)$ is calculated as 0.05.

Furthermore, for the AC cable, the effective inductance for each phase, $L_{\text{Cable,ph}}$, is calculated with equation (4.3), where D_{ph2ph} represents the center-to-center distance between phases [123]. In this cable, D_{ph2ph} is approximately 23.5 mm.

$$L_{\text{Cable,ph}} = \frac{\mu_0}{2\pi} \cdot l_{\text{Cable}} \cdot \ln\left(\frac{2D_{\text{ph2ph}}}{e^{-\frac{1}{4}} \cdot D_{\text{Tape}}}\right) \quad (4.3)$$

4.2.4 Cable Coolant

As previously mentioned, a superconductor is in the superconducting state when it is cooled down to a temperature lower than its critical temperature T_c. A coolant is used to provide such a condition. Considering that coolants are pumped from one side of the cable, their pressure and temperature is not constant along the cable; the pressure drops and the temperature rises [124], [125]. For the cable to operate safely, the coolant's temperature and pressure must remain within acceptable limits. Figure 4.4 shows the pressure-temperature diagram of nitrogen, whose liquid form is used as the cable coolant in this work [124].

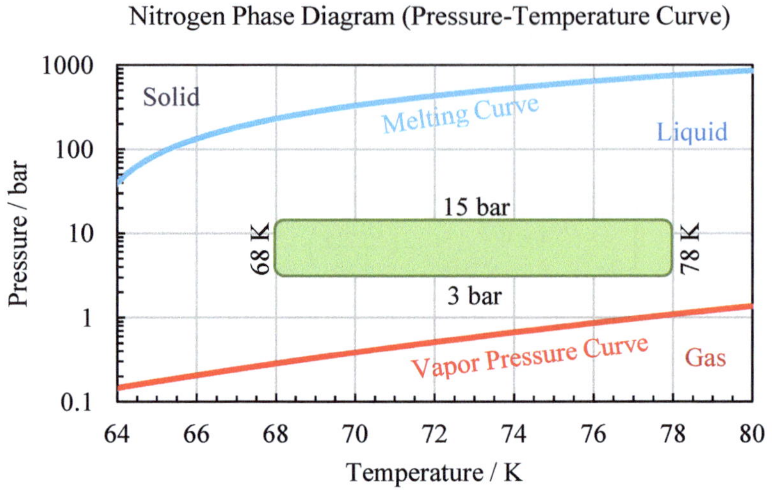

Figure 4.4: Nitrogen phase diagram (pressure-temperature) curve, where the limits for the phase change are shown, and the green area is the safe operating region for using it in liquid form (LN_2) [124].

Assuming that LN_2 is pumped at 68 K and 15 bar, a pressure drop higher than 12 bar or a temperature rise above 78 K poses a risk of LN_2 vaporizing. Due to the large density difference between liquid and gaseous nitrogen ($\rho_{\mathrm{LN_2}} > 130\rho_{\mathrm{GN_2}}$), vaporization can transform 1 liter of liquid nitrogen into tens of liters of gas, potentially leading to a loss of superconductivity, and deformation and insulation voltage breakdown of the cable [126]. Therefore, in the case of using LN_2 as a coolant, the safe operating range is recommended to be the green area in Figure 4.4 [124], [125]. Consequently, analyzing pressure drops and temperature rises along the cable is essential to ensure the coolant operates safely. In this work, these parameters are analyzed only for the DC cable coolant, however, the same approach can be applied to the AC cable.

4.2.4.1 Calculation of the pressure drop along the cable length

The following Darcy-Weisbach equation can be used to calculate coolant pressure drop along the cable [124], [125], [127].

$$\Delta p = \lambda_{\mathrm{Fld}} \cdot \frac{l_{\mathrm{Cable}}}{2\rho_{\mathrm{Fld}} \cdot d_{\mathrm{H,Fld}}} \cdot \left(\frac{\dot{m}_{\mathrm{Fld}}}{A_{\mathrm{Fld}}}\right)^2 \tag{4.4}$$

In this equation, ρ_{Fld}, A_{Fld}, and $\dot{m}_{\mathrm{Fld}}$ are the fluid density, area, and mass flow rate, respectively. The parameter $d_{\mathrm{H,Fld}}$ represents the fluid hydraulic diameter which is the difference between fluid duct outer and inner diameters. In the case of the DC cable considered in this work, this parameter is 10 mm. Moreover, in the above equation, the parameter λ_{Fld} is the friction coefficient. The following points that resulted from Equation (4.4) must be considered:

- The pressure drop in a cable increases as its length increases.

- Pumping the coolant at a higher mass flow rate results in a higher pressure drop.
- Using gases as coolants instead of liquid coolants, such as helium gas at 77 K, reduces the pressure drop because of the significantly smaller mass flow rates.
- Employing a less dense fluid, like liquid hydrogen (LH_2), compared to liquid nitrogen (LN_2), increases the pressure drop.
- A thicker coolant duct, meaning a larger hydraulic diameter $d_{\mathrm{H,Fld}}$, reduces the pressure drop.

To calculate λ_{Fld}, the model for corrugated tubes investigated by Kauder can be used [128]. This parameter is calculated with equation (4.5) from cooling fluid Reynolds number Re_{Fld} for a smooth tube (where the ratio of the tube groove depth over tube groove width is less than 0.2) based on the Karman-Nikuradse correlation [124]. To calculate the fluid Reynolds number Re_{Fld}, the procedure explained in the appendices Chapter 9, section A.1 must be followed [129].

$$\frac{1}{\sqrt{\lambda_{\mathrm{Fld}}}} = 2log\left(\frac{Re_{\mathrm{Fld}} \cdot \sqrt{\lambda_{\mathrm{Fld}}}}{2.51}\right), Re > 2300 \tag{4.5}$$

By solving this equality in equation (4.5), the friction coefficient λ_{Fld} is calculated as described in equation (4.6).

$$\lambda_{\mathrm{Fld}} = \frac{1}{4\omega^2\left(-\,log\left(\frac{251}{50Re_{\mathrm{Fld}}}\right)\right)} \tag{4.6}$$

In this equation, $\omega(z)$ is the Wright Omega function. Based on [128], for the duct type described above, equation (4.6) can be used when $Re_{\mathrm{Fld}}>5\times10^4$.

Figure 4.5 demonstrates the result of the LN_2 pressure drop analysis that is pumped at 15 bar and 77 K in the cable described in Figure 4.2 and Table 4.2. This figure confirms that this coolant will have a negligible pressure drop, and it remains in the safe operating region described in Figure 4.4.

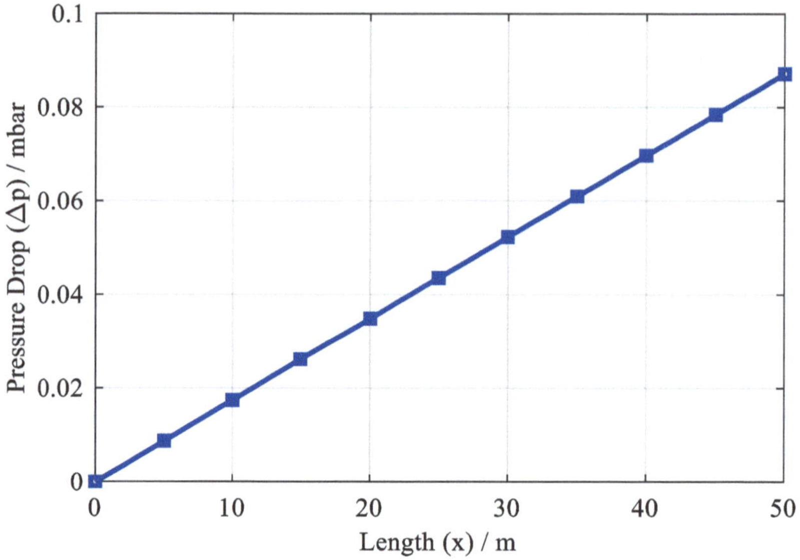

Figure 4.5: Pressure drop analysis results for LN_2, which is pumped at 15 bar and 77 K in the DC cable described in Figure 4.2 and Table 4.2, based on the LN_2 properties obtained from the NIST database [126]. This data is given in the appendices Chapter 9, section A.3, Table 9.2.

4.2.4.2 Calculation of the temperature rise along the cable length

Figure 4.6 shows a cooling fluid duct element at the distance x from the fluid pumping side of the cable and with a length of dx. The parameter dx is considered relatively small with no temperature gradient. In other words, the temperature is the same as $T_{\text{Fld,x}}$ at every point of this element.

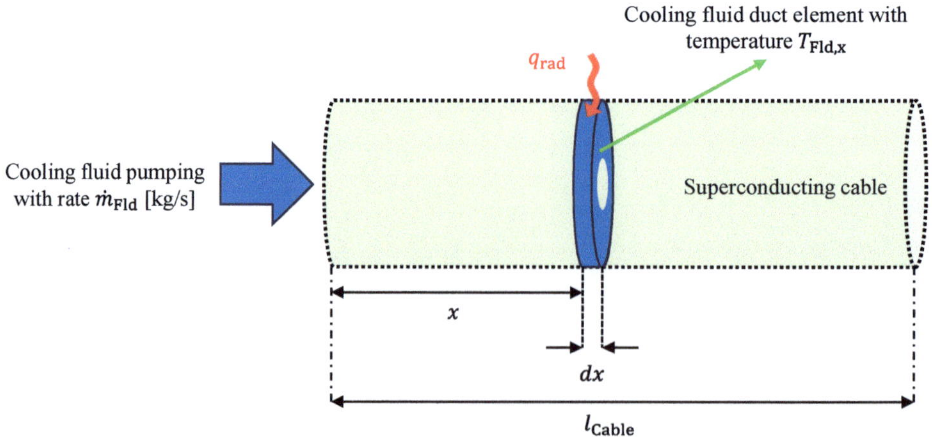

Figure 4.6: Illustration of a coolant duct element with a significantly small length of dx with a temperature $T_{\text{Fld,x}}$ (no temperature gradient).

Considering a radiated heat input from outside as q_{rad} (constant along the cable, considered 1 W/m in this work) and $c_{\mathrm{p,Fld}}$ as cooling fluid's specific heat capacity, the heat balance equation can be written as (4.7).

$$\dot{m}_{\mathrm{Fld}} \cdot c_{\mathrm{p,Fld}} \cdot \frac{dT_{\mathrm{Fld,x}}}{dx} = q_{\mathrm{rad}} \tag{4.7}$$

Solving this equation results in calculating the fluid temperature at a distance x from the pumping side ($T_{\mathrm{Fld,x}}$). This is shown in equation (4.8), where $T_{\mathrm{Fld,x_0}}$ is the coolant temperature at the beginning of the cable.

$$T_{\mathrm{Fld,x}} = \frac{q_{\mathrm{rad}}}{\dot{m}_{\mathrm{Fld}} \cdot c_{\mathrm{p,Fld}}} \cdot x + T_{\mathrm{Fld,x_0}} \tag{4.8}$$

In addition to the pressure drop, equation (4.8) shows another criterion for the cable design and coolant selection. This equation highlights the following points:

- Considering a coolant duct with a larger outer diameter will lead to a greater temperature rise along the cable length.
- Using a gaseous coolant, which is pumped at a significantly lower mass flow rate, can considerably raise the temperature along the cable length.
- Selecting a coolant with a higher heat capacity, such as LH_2 instead of LN_2, reduces the temperature rise.

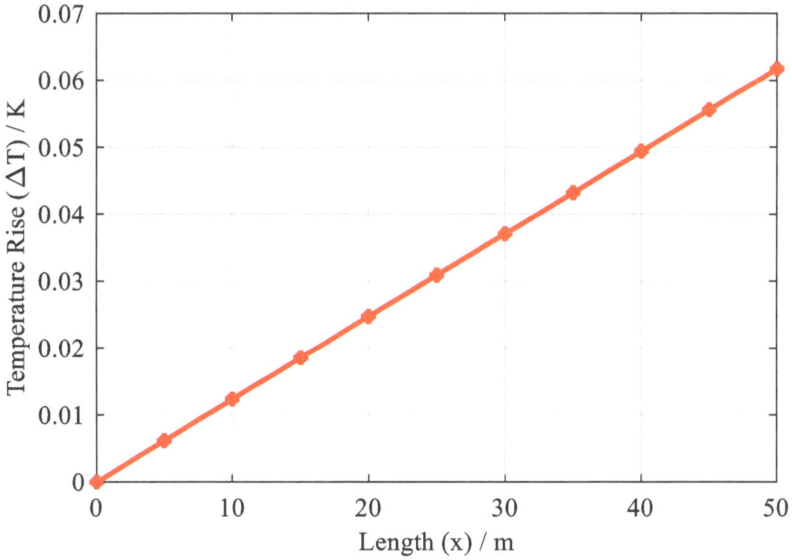

Figure 4.7: Temperature rise analysis results for LN_2, which is pumped at 15 bar and 77 K in the DC cable described in Figure 4.2 and Table 4.2, based on the LN_2 properties obtained from the NIST database [126]. This data is given in the appendices Chapter 9, section A.3, Table 9.2.

Figure 4.7 presents the analysis result of the LN_2 temperature rise along the cable length for the cable described in Figure 4.2 and Table 4.2. This figure indicates that the LN_2 will experience a negligible temperature increase of 0.06 K, which falls within the safe operating range for LN_2, as described in Figure 4.4. This case assumes an adiabatic condition for the coolant. In reality since coolant has a heat convection with other layers, its temperature rise will be even lower.

4.3 Electrical Characteristic

A superconducting DC cable can be represented by a resistive-inductive impedance in an electrical system. The equivalent electrical circuit of a superconducting cable is shown in Figure 4.8.

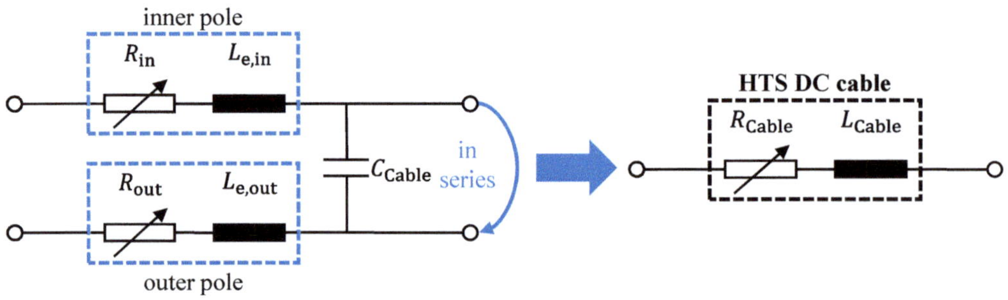

Figure 4.8: Electrical equivalent circuit of a bipolar coaxial superconducting DC cable, described in Figure 4.2 and Table 4.2.

In this figure, $L_{\text{e,in}}$ and $L_{\text{e,out}}$ denote the effective inductances of each pole as explained in the sub-chapter 4.2.3. The cable capacitance C_{Cable} represents the capacitive charges between the poles. As the voltage rating in the cable considered in this work is relatively low, and due to the existence of the PPLP for electrical insulation between poles, the cable capacitance is neglected. Since the two poles are in series, the total cable resistance R_{Cable} and inductance L_{Cable} can be calculated based on the series impedance rule. Calculation of the cable inductance L_{Cable} was explained in the sub-chapter 4.2.3.

Moreover, R_{in} and R_{out} are the non-linear resistances of the inner and outer poles, respectively, due to the existence of the superconductor. The cable resistance R_{Cable} consists of the HTS tapes used in the cable, and the core/shield copper layers parallel to the tapes. This follows the same principle as RSFCL explained in the sub-chapter 3.4. The detailed equivalent circuit of the cable resistance is shown in Figure 4.9.

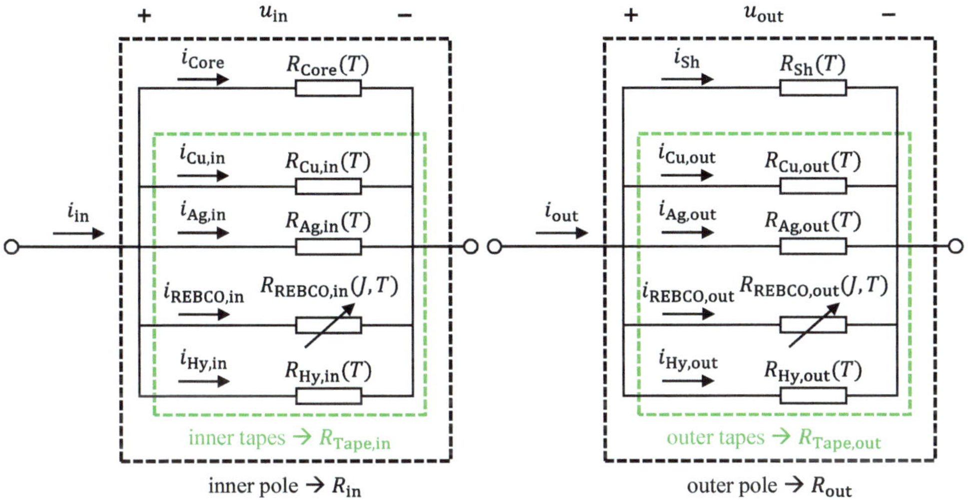

Figure 4.9: Electrical equivalent circuit of a bipolar coaxial superconducting DC cable resistance, considering the resistances of all the layers and sub-layers of the HTS tapes, with the principle explained in the sub-chapter 3.4. In this figure, the indices "in" and "out" refer to the inner and outer poles. In this bipolar DC structure, the poles currents, i_{in} and i_{out}, are equal.

In Figure 4.9, the resistance of each sub-layer in the tapes represents its total resistance in all parallel tapes. For instance, $R_{REBCO,in}$ indicates the total resistance of REBCO sub-layers in all inner pole HTS tapes. In this analysis, it is assumed that all tapes have identical properties, temperatures, and currents. The parameters $R_{Tapes,in}$ and $R_{Tapes,out}$ denote the equivalent resistances of the HTS tapes in the inner and outer poles, respectively. These parameters can be calculated using the parallel resistors law, similar to what was explained for the RSFCL in the sub-chapter 3.4, equation (3.7). Likewise, to determine the total resistance of each pole from $R_{Tapes,in}$, $R_{Tapes,out}$, and shield/copper resistances, R_{Core} and R_{Sh}, the same law can be used. The current flowing through each layer is also illustrated; For instance, the current through the REBCO layers in the inner tapes is labelled $i_{REBCO,in}$. Furthermore, the voltage on the two poles are shown as u_{in} and u_{out}.

As explained in the sub-chapter 3.2, the resistance behavior of REBCO is non-linear with current density J and temperature T, while the resistance of other layers varies only with temperature and can be calculated using equation (3.6) in Chapter 3. The specific data to calculate these resistances are given in appendices Chapter 9, section A.1, Table 9.1. Finally, the total cable resistance is calculated using equation (4.9).

$$R_{Cable} = R_{in} + R_{out} \tag{4.9}$$

Following the same principle as the DC cable, the equivalent circuit of the three-phase HTS AC cable is illustrated in Figure 4.10. In this figure, $R_{Cable,abc}$ denote the three-phase cable resistance, and $L_{Cable,abc}$ represent its inductance. With the same principle as DC cable, the cable capacitance in each phase is neglected due to the insulation, and low voltage rating and frequency.

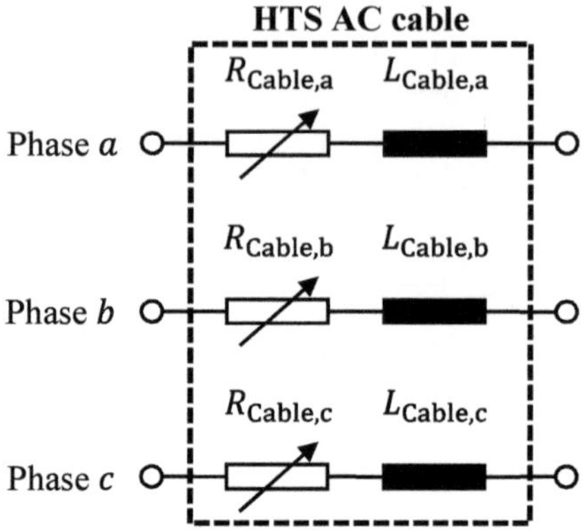

Figure 4.10: Electrical equivalent circuit of a three-phase superconducting AC cable, described in Figure 4.3 and the geometry provided in Table 4.3.

The resistance of each phase, $R_{\text{Cable,ph}}$ is non-linear because the REBCO layers resistances in the HTS tapes change non-linearly with current and temperature. The equivalent circuit of the phases resistance is depicted in Figure 4.11, with the same principles as described in Chapter 4, sub-chapter 4.3. The total tapes resistance per phase, $R_{\text{Tape,ph}}$, and phase resistance $R_{\text{Cable,ph}}$ are calculated using parallel resistors law, considering the same properties for all HTS tapes. The parameters needed to calculate the resistance of all the layers are provided in the appendices Chapter 9, section A.1, Table 9.1.

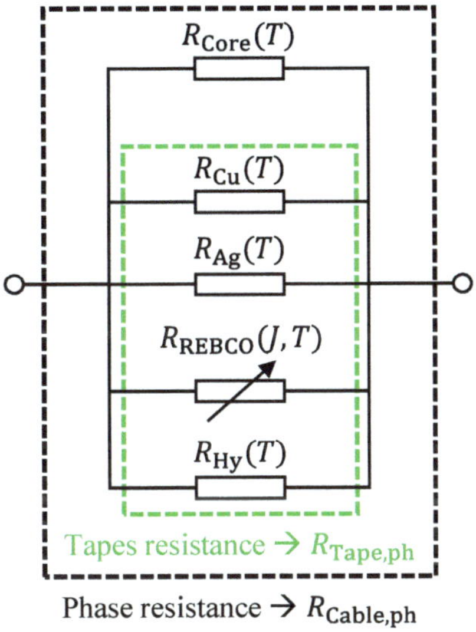

Figure 4.11: Electrical equivalent circuit of a superconducting AC cable phase resistance, considering the resistances of all the layers and sub-layers of the HTS tapes, with the principle explained in the sub-chapter 3.4.

4.4 Non-linear, Adiabatic, Lumped-parameter Model

The lumped-parameter method applied to the cable follows the same principle outlined for the RSFCL in Chapter 3, sub-chapters 3.5 and 3.6. In this approach, each layer has a single temperature and resistance. This method only uses adiabatic conditions, meaning no heat transfer between layers is considered. Hence, the temperature of a layer varies only with the power loss within that layer. This temperature is determined using equation (3.11) in Chapter 3. The process for determining the various parameters of the cable layers is based on the algorithm described in the sub-chapter 3.6, Figure 3.6.

4.5 Non-linear, Non-adiabatic, One-dimensional, Finite-difference-method-based Model

A more advanced model than the lumped-parameter is needed to model the cable in a non-adiabatic environment. A non-adiabatic model includes the heat transfer between the cable layers. For this purpose, the finite-difference method (FDM) is used to model the thermal characteristic of the cable [125], [130], [131]. With this method, the cable can be modeled one-dimensionally or two-dimensionally. This sub-chapter presents the one-dimensional (1-D), FDM-based modeling of the cable.

In the 1-D modeling of the cable using FDM, the heat transfer of the cable only along its cross-section (radial direction) is considered. Therefore, the cable cross-section is divided into multiple elements with a small thickness indicated by Δr. Figure 4.12 shows a part of the cable cross-section with the finite-difference elements.

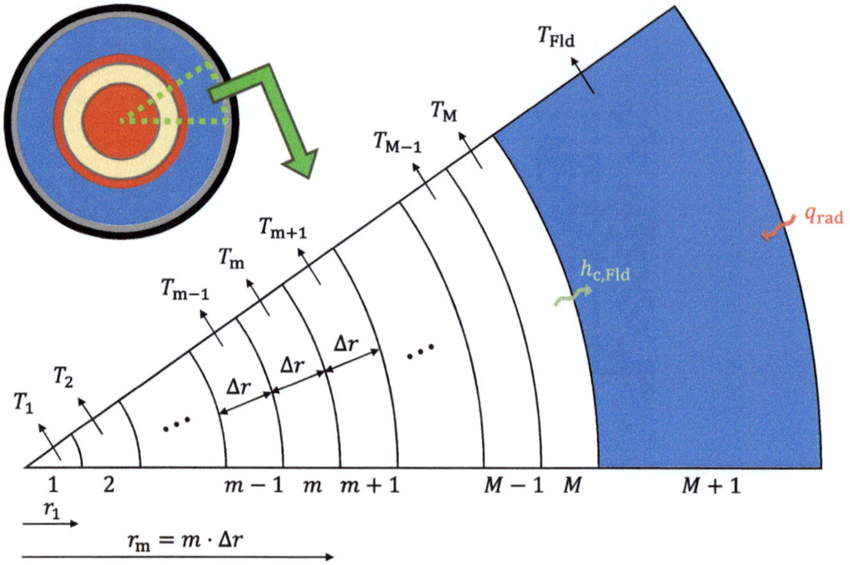

Figure 4.12: Illustration of an HTS cable section with applied finite difference method (FDM), where solid layers are divided into M elements, while coolant has one element with a temperature of T_{Fld}. Each solid element has a thickness of Δr, with T_{m} representing the temperature of mth element and r_{m} denoting the distance of mth element from the cable radial center. Additionally, q_{rad} represents the radiated heat from outside, assumed to be 1 W/m in this work, and $h_{\mathrm{c,Fld}}$ indicates the cooling fluid convective heat coefficient.

The parameter Δr and consequently total number of FDM elements must be chosen with a compromise between the computational effort and the accuracy of the results. If the number of elements is too high, the time to simulate the model would be significantly prolonged, though the results have remarkably high accuracy. Conversely, if the number of the FDM elements is too low, the results might be inaccurate. Since no temperature gradient is considered within the HTS tapes due to their thinness, the parameter Δr is set equal to the tapes thickness, d_{Tape}, as specified in the section 4.2.1, Table 4.1. Table 4.4 shows the 1-D FDM parameters applied to the cable.

Table 4.4: Specifications of the 1-D finite difference method (FDM) applied to the HTS DC cable with the structure described in Figure 4.2 and Table 4.2.

Parameter	Symbol	Value
Thickness of the Radial Elements in the Solid Layers	Δr	120 µm
Total Number of the Elements in the Solid Layers	M	74
Number of the Radial Elements for the Cooling Fluid	$N_{\mathrm{r,Fld}}$	1
Total Number of the Radial Elements	N_{r}	75

With this data, the area and volume of a certain element m, denoted by A_{m} and V_{m} can be determined using equations (4.10) and (4.11).

$$A_{\mathrm{m}} = \pi \cdot (2m - 1) \cdot (\Delta r)^2 \tag{4.10}$$

$$V_{\mathrm{m}} = A_{\mathrm{m}} \cdot l_{\mathrm{Cable}} \tag{4.11}$$

The temperature of each element can be determined using FDM based on heat transfer principles. Heat transfer in the cable can occur in two states: stationary and transient. The stationary state refers to the cable operating under nominal conditions, where all parameters are stable at their nominal values. Modeling the cable in this state is essential as it provides the initial condition for further analysis. In contrast, the transient state involves analyzing temperature changes over time. Transient modeling is used to describe the cable's behavior during dynamic events, such as sudden load changes caused by short circuits in the cable or the network where it is installed. The procedure for calculating the temperature in the stationary state is explained first, as it is used to determine the initial temperature of the cable. Once the stationary state is established, the analysis moves on to the transient state, where temperature variations over time are examined.

4.5.1 Stationary State

The general 1-dimensional, stationary heat transfer equation is shown in equation (4.12) [130], [131].

$$\frac{d^2T}{dr^2} + \frac{1}{r} \cdot \frac{dT}{dr} + \frac{g}{k} = 0 \tag{4.12}$$

In this equation, the parameters T, g, and k represent temperature, generated power, and thermal conductivity, respectively. For a certain element m within the cable, this equation can be written in FDM form. By formulating the equation for one specific element, they can then be extended to all other elements. Equation (4.13) describes the heat transfer equation of the element m^{th} in the FDM form, incorporating the boundary conditions with its adjacent elements, m-1 and m+1. The properties of all the materials are provided in the appendices Chapter 9, section A.1.

$$\left[\left(\frac{1}{2m}-1\right)\cdot\left(\frac{2}{1+\eta_{m,m-1}}\right)\right]\cdot T_{m-1}$$
$$+\left[2+\left(\frac{1}{2m}-1\right)\cdot\left(\frac{\eta_{m,m-1}-1}{\eta_{m,m-1}+1}\right)+\left(\frac{1}{2m}+1\right)\cdot\left(\frac{\eta_{m+1,m}-1}{\eta_{m+1,m}+1}\right)\right]\cdot T_m \quad (4.13)$$
$$+\left[\left(\frac{1}{2m}+1\right)\cdot\left(\frac{-2\eta_{m+1,m}}{1+\eta_{m+1,m}}\right)\right]\cdot T_{m+1}=\frac{(\Delta r)^2}{k_m}\cdot g_m$$

In this equation, g_m denotes the generated power in the element m, which is calculated with equation (4.14), where R_m and I_m are the resistance and flowing current in this element. Moreover, V_m is its volume which its calculation was described in equation (4.11).

$$g_m=\frac{R_m\cdot I_m^2}{V_m} \quad (4.14)$$

Moreover, $\eta_{m,m-1}$ and $\eta_{m+1,m}$ represent the thermal conductivity ratios between the mth element and its neighboring elements, as described in (4.15). In this equation, k_{m-1}, k_m, and k_{m+1} are the thermal conductivities of the elements m-1, m, and m+1, respectively.

$$\eta_{m,m-1}=\frac{k_m}{k_{m-1}}\,,\eta_{m+1,m}=\frac{k_{m+1}}{k_m} \quad (4.15)$$

The equation (4.13) is efficient as it eliminates the need to implement different equations for the elements. Moreover, there is no need to check for different thermal conductivities since this equation considers the boundary conditions with both adjacent elements.

There are particular elements within solid layers with specific heat transfer equations: first solid element (m=1), and last solid element (m=M). The FDM forms of the 1-D heat transfer equation for element m=1 and m=M are shown in (4.16) and (4.17), respectively.

$$4T_1-4T_2=\frac{(\Delta r)^2}{k_1}\cdot g_1 \quad (4.16)$$

$$-2T_{M-1}+\left[2+\varphi\cdot\left(\frac{1}{2M}+1\right)\right]T_M+\left[-\varphi\cdot\left(\frac{1}{2M}+1\right)\right]\cdot T_{Fld}$$
$$=\frac{(\Delta r)^2}{k_M}\cdot g_M \quad (4.17)$$

In equation (4.17), the parameter φ is a self-built parameter which is given in (4.18), where k_M is the thermal conductivity of last solid element, m=M.

$$\varphi = \frac{2\Delta r \cdot h_{\mathrm{c,Fld}}}{k_{\mathrm{M}}} \tag{4.18}$$

The parameter $h_{\mathrm{c,Fld}}$ represents the cooling fluid (LN_2) convective heat coefficient which depends on multiple parameters, mostly related to the cable geometry and properties of each specific fluid itself. To calculate this parameter, the procedure described in the appendices Chapter 9, section A.1 must be followed. In the stationary state simulated by 1-D model, the temperature of the last element, the cooling fluid, is set at 77 K.

The equations for calculating the temperature in all FDM elements are now found. Since the temperature of each element also depends on the temperature of one or two adjoining elements, a mathematical approach is needed to calculate all these temperatures. These equations can be written in a matrix form as follows:

$$A \times T = B \tag{4.19}$$

In this equation, T is the temperature matrix that contains the temperature of all FDM elements, and its size is $N_{\mathrm{r}} \times 1$. Additionally, in equation (4.19), A is the known matrix of coefficients in the heat transfer equations (On the left side of the equations), and B is the known matrix of power generated in the elements (On the right side of the equations). The sizes of matrix A and B are $N_{\mathrm{r}} \times N_{\mathrm{r}}$ and $N_{\mathrm{r}} \times 1$, respectively.

The matrices T and B are shown in equation (4.20), and the matrix A is presented in equation (4.21). In the matrix A, the array $\mathrm{A}_{\mathrm{i,j}}$ represents the contribution of element $m = j$ (i.e., T_{j}) to the calculation of the temperature of element $m = i$ (i.e., T_{i}). For example, the array $A_{\mathrm{m,m-1}}$ is the coefficient of $T_{\mathrm{m-1}}$ in the heat transfer equation of element m (as described in equation (4.13)). Given the large sizes of these matrices, only specific arrays are depicted, while others follow the same procedure. Moreover, in matrix A, the arrays related to heat transfer of a certain element m and last solid element $m = M$, based on equations (4.13) and (4.17) are shown.

$$T = \begin{bmatrix} T_1 \\ T_2 \\ \vdots \\ T_{\mathrm{m-1}} \\ T_{\mathrm{m}} \\ T_{\mathrm{m+1}} \\ \vdots \\ T_{\mathrm{M-1}} \\ T_{\mathrm{M}} \\ T_{\mathrm{Fld}} \end{bmatrix} \quad , \quad \mathrm{B} = \begin{bmatrix} \frac{(\Delta r)^2}{k_1} \cdot g_1 \\ \frac{(\Delta r)^2}{k_2} \cdot g_2 \\ \vdots \\ \frac{(\Delta r)^2}{k_{\mathrm{m-1}}} \cdot g_{\mathrm{m-1}} \\ \frac{(\Delta r)^2}{k_{\mathrm{m}}} \cdot g_{\mathrm{m}} \\ \frac{(\Delta r)^2}{k_{\mathrm{m+1}}} \cdot g_{\mathrm{m+1}} \\ \vdots \\ \frac{(\Delta r)^2}{k_{\mathrm{M-1}}} \cdot g_{\mathrm{M-1}} \\ \frac{(\Delta r)^2}{k_{\mathrm{M}}} \cdot g_{\mathrm{M}} \\ 77\,\mathrm{K} \end{bmatrix} \tag{4.20}$$

$$\mathrm{A} = \begin{bmatrix} 4 & -4 & 0 & \cdots & 0 & 0 & 0 & \cdots & 0 & 0 \\ -\frac{3}{4} & 2 & -\frac{5}{4} & \cdots & 0 & 0 & 0 & \cdots & 0 & 0 \\ \vdots & \vdots & \vdots & \ddots & \vdots & \vdots & \vdots & \ddots & \vdots & \vdots \\ 0 & 0 & 0 & \cdots & A_{\mathrm{m-1,m-1}} & A_{\mathrm{m-1,m}} & 0 & \cdots & 0 & 0 \\ 0 & 0 & 0 & \cdots & A_{\mathrm{m,m-1}} & A_{\mathrm{m,m}} & A_{\mathrm{m,m+1}} & \cdots & 0 & 0 \\ 0 & 0 & 0 & \cdots & 0 & A_{\mathrm{m+1,m}} & A_{\mathrm{m+1,m+1}} & \cdots & 0 & 0 \\ \vdots & \vdots & \vdots & \ddots & \vdots & \vdots & \vdots & \ddots & \vdots & \vdots \\ 0 & 0 & 0 & \cdots & 0 & 0 & 0 & \cdots & A_{\mathrm{M-1,M}} & 0 \\ 0 & 0 & 0 & \cdots & 0 & 0 & 0 & \cdots & A_{\mathrm{M,M}} & A_{\mathrm{M,N_r}} \\ 0 & 0 & 0 & \cdots & 0 & 0 & 0 & \cdots & 0 & 1 \end{bmatrix}$$

$$A_{\mathrm{m,m-1}} = \left(\frac{1}{2m} - 1\right) \cdot \left(\frac{2}{1 + \eta_{\mathrm{m,m-1}}}\right) \tag{4.21}$$

$$A_{\mathrm{m,m}} = 2 + \left(\frac{1}{2m} - 1\right) \cdot \left(\frac{\eta_{\mathrm{m,m-1}} - 1}{\eta_{\mathrm{m,m-1}} + 1}\right) + \left(\frac{1}{2m} + 1\right) \cdot \left(\frac{\eta_{\mathrm{m+1,m}} - 1}{\eta_{\mathrm{m+1,m}} + 1}\right)$$

$$A_{\mathrm{m,m+1}} = \left(\frac{1}{2m} + 1\right) \cdot \left(\frac{-2\eta_{\mathrm{m+1,m}}}{1 + \eta_{\mathrm{m+1,m}}}\right)$$

$$A_{\mathrm{M,M-1}} = -2 \quad , \quad A_{\mathrm{M,M}} = 2 + \varphi \cdot \left(\frac{1}{2M} + 1\right), \qquad A_{\mathrm{M,N_r}} = -\varphi \cdot \left(\frac{1}{2M} + 1\right)$$

By building all the matrices in equation (4.19), the temperature matrix T in the stationary regime can be calculated with equation (4.22).

$$T = A^{-1} \times B \tag{4.22}$$

4.5.2 Transient State

The temperature calculation in the 1-D transient model follows a similar process to that of the stationary model, with the key difference being the inclusion of temperature variation over time t. The transient heat transfer in the FDM elements shown in Figure 4.12 is based on the equation (4.23) [130], [131]. Unlike the 1-D stationary heat transfer described by equation (4.12), where the right-hand side equals zero, in the transient model, the right side of the equation accounts for the time-dependent changes in generated power within the elements. As a result, temperature variation over time is considered alongside heat transfer.

$$\frac{d^2T}{dr^2} + \frac{1}{r} \cdot \frac{dT}{dr} + \frac{g}{k} = \frac{\rho \cdot c_\mathrm{p}}{k} \cdot \frac{dT}{dt} \tag{4.23}$$

In this equation, ρ represents the density and c_p is the specific heat capacity of the material. Similar to the FDM form used in the stationary heat transfer model, the 1-D heat transfer equation for the transient state can also be derived. This work employs the implicit method to compute temperatures in the 1-D transient model, meaning that the temperatures at each new time step must be calculated. For a certain element m in solid layers, the 1-D, transient heat transfer equation is as follows:

$$\begin{aligned} &\left[\left(\frac{1}{2m} - 1\right) \cdot \left(\frac{2}{\eta_\mathrm{m,m-1} + 1}\right)\right] \cdot T_\mathrm{m-1}^\mathrm{z+1} \\ &+ \left[\psi_\mathrm{m} + 2 + \left(\frac{1}{2m} - 1\right) \cdot \left(\frac{\eta_\mathrm{m,m-1} - 1}{\eta_\mathrm{m,m-1} + 1}\right) + \left(\frac{1}{2m} + 1\right) \cdot \left(\frac{\eta_\mathrm{m+1,m} - 1}{\eta_\mathrm{m+1,m} + 1}\right)\right] \\ &\cdot \mathrm{T}_\mathrm{m}^\mathrm{z+1} \\ &+ \left[\left(\frac{1}{2m} + 1\right) \cdot \left(\frac{-2\eta_\mathrm{m+1,m}}{\eta_\mathrm{m+1,m} + 1}\right)\right] \cdot T_\mathrm{m+1}^\mathrm{z+1} = \psi_\mathrm{m} \cdot T_\mathrm{m}^\mathrm{z} + \frac{(\Delta r)^2}{k_\mathrm{m}} \cdot g_\mathrm{m}^\mathrm{z+1} \end{aligned} \tag{4.24}$$

In this equation, the left-hand side contains the unknown temperatures and their corresponding coefficients, which are to be calculated for the new time step, $z+1$. The right-hand side includes known parameters, such as the temperature from the previous time step (i.e., T_m^z) and the power generated in each element at the new time step $g_\mathrm{m}^\mathrm{z+1}$. Additionally, ψ_m in equation (4.24) is defined like equation (4.25) from element m density, specific heat capacity, and thermal conductivity which are represented respectively with ρ_m, $c_\mathrm{p,m}$, and k_m. In this equation, Δt denotes the time-step interval. The properties of all the materials are provided in the appendices Chapter 9, section A.1.

$$\psi_\mathrm{m} = \frac{(\Delta r)^2 \cdot \rho_\mathrm{m} \cdot c_\mathrm{p,m}}{k_\mathrm{m} \cdot \Delta t} \tag{4.25}$$

The procedure to find the heat transfer equations for other radial elements in solid layers follows the same approach as explained above and the stationary state. These equations can be found in the appendices Chapter 9, section A.1.

In the transient model, the temperature of the cooling fluid also varies over time. Using the temperature of the last radial element in solid layers, T_{M}, along with the fluid convective heat coefficient $h_{\mathrm{c,Fld}}$, the fluid temperature can be calculated using equation (4.26). In this equation, ρ_{Fld} represents the fluid density, and $c_{\mathrm{p,Fld}}$ is the specific heat capacity of the fluid. Additionally, the parameter $Pe_{\mathrm{Fld},in}$ represents the fluid duct inner diameter ($=\pi \cdot d_{\mathrm{Sh}}$ in the cable considered in this work).

$$\begin{aligned} -h_{\mathrm{c,Fld}} \cdot Pe_{\mathrm{Fld,in}} \cdot T_{\mathrm{M}}^{\mathrm{z}+1} + \left[\psi_{\mathrm{Fld}} + h_{\mathrm{c,Fld}} \cdot Pe_{\mathrm{Fld,in}}\right] \cdot T_{\mathrm{Fld}}^{\mathrm{z}+1} \\ = \psi_{\mathrm{Fld}} \cdot T_{\mathrm{Fld}}^{\mathrm{z}} + q_{\mathrm{rad}} \end{aligned} \tag{4.26}$$

In this equation, $T_{\mathrm{M}}^{\mathrm{z}+1}$ and $T_{\mathrm{Fld}}^{\mathrm{z}+1}$ are the temperature of the last solid element $m=M$ and cooling fluid in the new time-step, which need to be calculated. The parameter $T_{\mathrm{Fld}}^{\mathrm{z}}$ is the fluid temperature in the previous time-step. The parameter ψ_{Fld} is a self-built parameter as described in equation (4.27).

$$\psi_{\mathrm{Fld}} = \frac{\rho_{\mathrm{Fld}} \cdot c_{\mathrm{p,Fld}} \cdot A_{\mathrm{Fld}}}{\Delta t} \tag{4.27}$$

By having all the equations, the same approach as the stationary model can be used to calculate the temperature in the transient model. In this equation, matrix θ is built to simplify the process. Therefore, the matrix form of the equations is built, as shown in equation (4.28). The approach of building these matrices is described in the appendices Chapter 9, section A.1.

$$\mathrm{A} \times T^{\mathrm{z}+1} = \psi \cdot T^{\mathrm{z}} + \mathrm{B} = \theta^{\mathrm{z}+1} \tag{4.28}$$

The temperature matrix in the new time-step $T^{\mathrm{z}+1}$ is then calculated with equation (4.29).

$$T^{\mathrm{z}+1} = \mathrm{A}^{-1} \times \theta^{\mathrm{z}+1} \tag{4.29}$$

4.6 Non-linear, Non-adiabatic, Two-dimensional, Finite-difference-method-based Model

In the 2-dimensional modeling of the cable, the temperature in both radial and axial directions is calculated. In this model, in addition to the cable cross-section (depicted in Figure 4.12), the cable length (x-direction) is also divided into N_{x} number of elements with the same length defined as Δx. Hence, the temperature of the cable layers along its length can be calculated with the 2-D model. This is crucial, especially in long-length cables where the coolant temperature can rise along the length, as discussed in section 4.2.4 of this chapter. In this cable, cooling fluid, LN_2 at 77 K, is pumped from one side (n=1) and leaves the cable towards DC/AC inverter on the other side at the end of the cable (n=N_x). The finite difference demonstration of the cable along its length is depicted in Figure 4.13.

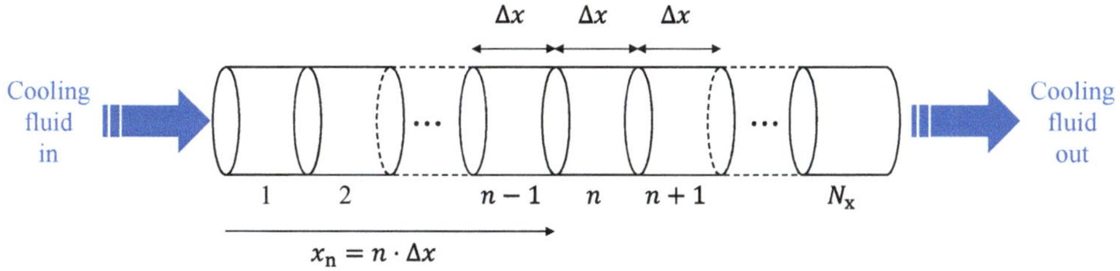

Figure 4.13: Illustration of an HTS cable axial direction with applied finite difference method (FDM), where its length is divided into N_x elements with a length of Δx. A certain element in this schematic is reprsented with a number n, and its distance from the starting point of the cable, where the cooling fluid is pumped, is denoted by x_n.

Similar to the values of Δr in the radial direction, choosing a value for Δx is a compromise between computational effort and accuracy. In this work, the length of each element Δx is chosen as 1 m. Therefore, the total number of axial elements N_x is calculated from the cable length l_{Cable}. Table 4.5 shows the 2-D FDM parameters applied to the cable.

Table 4.5: Specifications of the axial elements in the 2-D finite differene method (FDM) applied to the HTS DC cable with the structure described in Figure 4.2 and Table 4.2.

Parameter	Symbol	Value
Length of the Axial Elements	Δx	1 m
Total Number of the Axial Elements	N_x	50

If A_m is the area of radial element m (calculated with equation (4.10)), the volume of this element in the axial element n, $V_{m,n}$, is determined as follows:

$$V_{m,n} = A_m \cdot \Delta x \tag{4.30}$$

Similar to the 1-D model, the 2-D modeling of the cable in both stationary and transient conditions must be studied.

4.6.1 Stationary State

The 2-dimensional heat transfer equation in the stationary regime is given in equation (4.31) [130], [131].

$$\frac{d^2T}{dr^2}+\frac{1}{r}\cdot\frac{dT}{dr}+\frac{d^2T}{dx^2}+\frac{g}{k}=0 \tag{4.31}$$

As seen, the difference between 1-D, as described in equation (4.12), and 2-D models is the additional term d^2T/dx^2 which describes the heat transfer in x-direction (axial direction). To solve this equation using FDM, the temperature of the radial elements in each axial element is calculated, while the cooling fluid temperature in an axial element is updated based on the heat transfer with its adjacent axial elements. Writing the FDM form of this equation for a certain radial element m in the axial element n, including the boundary conditions with its adjoining elements based on equation (4.13) results in equation (4.32).

$$\begin{aligned}&\left[\left(\frac{1}{2m}-1\right)\cdot\left(\frac{2}{1+\eta_{\mathrm{m,m-1}}}\right)\right]\cdot T_{\mathrm{m-1,n}}\\&+\left[2+\left(\frac{1}{2m}-1\right)\cdot\left(\frac{\eta_{\mathrm{m,m-1}}-1}{\eta_{\mathrm{m,m-1}}+1}\right)+\left(\frac{1}{2m}+1\right)\cdot\left(\frac{\eta_{\mathrm{m+1,m}}-1}{\eta_{\mathrm{m+1,m}}+1}\right)\right]\cdot T_{\mathrm{m,n}}\\&+\left[\left(\frac{1}{2m}+1\right)\cdot\left(\frac{-2\eta_{\mathrm{m+1,m}}}{1+\eta_{\mathrm{m+1,m}}}\right)\right]\cdot T_{\mathrm{m+1,n}}=\frac{(\Delta r)^2}{k_{\mathrm{m}}}\cdot g_{\mathrm{m,n}}\end{aligned} \tag{4.32}$$

In this equation, the generated power in the radial element m in the axial element n, denoted $g_{\mathrm{m,n}}$, is determined with equation (4.33), where $R_{\mathrm{m,n}}$ and $I_{\mathrm{m,n}}$ are the resistance and flowing current in this element. The parameter $V_{\mathrm{m,n}}$ represents the volume of this element which is determined with equation (4.30).

$$g_{\mathrm{m,n}}=\frac{R_{\mathrm{m,n}}\cdot I_{\mathrm{m,n}}^2}{V_{\mathrm{m,n}}} \tag{4.33}$$

The difference between equations (4.13) and (4.32) is that in the 2-D model, the temperature of a radial element m in the axial element n is considered (i.e., $T_{\mathrm{m,n}}$ instead of T_{m}). The main difference between 1-D and 2-D, stationary models lies in the heat transfer of the cooling fluid. While in the 1-D model, the coolant temperature was considered constant, in the 2-D model, its variation along the length is considered. The heat transfer equation of the cooling fluid in the 2-D, stationary model is shown below:

$$\begin{aligned}&\left[-h_{\mathrm{c,Fld}}\cdot Pe_{\mathrm{Fld,in}}\right]\cdot T_{\mathrm{M,n}}+\left[h_{\mathrm{c,Fld}}\cdot Pe_{\mathrm{Fld}}+\frac{\dot{m}_{\mathrm{Fld}}\cdot c_{\mathrm{p,Fld}}}{\Delta x}\right]\cdot T_{\mathrm{Fld,n}}\\&=\left[\frac{\dot{m}_{\mathrm{Fld}}\cdot c_{\mathrm{p,Fld}}}{\Delta x}\right]\cdot T_{\mathrm{Fld,n-1}}+q_{\mathrm{rad}}\end{aligned} \tag{4.34}$$

In this equation, $T_{\mathrm{Fld,n}}$ represents the cooling fluid temperature at axial element n, while $T_{\mathrm{Fld,n-1}}$ denotes its temperature in the element before (closer to the beginning of the cable). The term $T_{\mathrm{Fld,n-1}}$ is on the right side of the equation (known side) since the temperature calculation is done radially for each axial element starting from $n = 1$, so the fluid temperature in the previous axial element is already calculated. Therefore, in this model, the temperature calculation starts from the first axial element, and then, the temperature would be calculated for the following elements with the same procedure. It needs to be mentioned that the fluid temperature at the first axial element n=1 is known and given as liquid nitrogen temperature, 77 K. The equations for other elements are given the appendices Chapter 9, section A.1.

Finally, all the equations are found, and the temperature matrix is calculated with the same approach as 1-D model. This is shown in equation (4.35).

$$\mathrm{A_n} \times T_\mathrm{n} = \mathrm{B_n} \Rightarrow T_\mathrm{n} = \mathrm{A_n^{-1}} \times \mathrm{B_n} \tag{4.35}$$

In this equation, T_n is the temperature of radial elements in axial element n. Subsequently, the matrix A_n is the coefficient matrix for this temperature matrix, and B_n is the known matrix for elements in the radial direction in axial element n. These matrices can be found based on the equations (4.32) and (4.34). These matrices are detailed in the appendices Chapter 9, section A.1.

The temperature calculation algorithm in the 2-D stationary model is shown in Figure 4.14. With this approach, by calculating the radial temperature in all axial elements, the main temperature matrix T can be found. This matrix has N_x number of columns where each corresponds to an axial element (Axial element $n \Leftrightarrow$ Column n in matrix T). Therefore, the matrix T is built by placing the calculated radial temperature matrix in each axial element in the corresponding column. This is shown in equation (4.36), where the array $T_\mathrm{m,n}$ represents the temperature of the radial element m in the axial element.

$$T = \left[[T_1] \quad [T_2] \quad \cdots \quad [T_\mathrm{n-1}] \quad [T_\mathrm{n}] \quad [T_\mathrm{n+1}] \quad \cdots \quad [T_{\mathrm{N_x}}]\right] \tag{4.36}$$

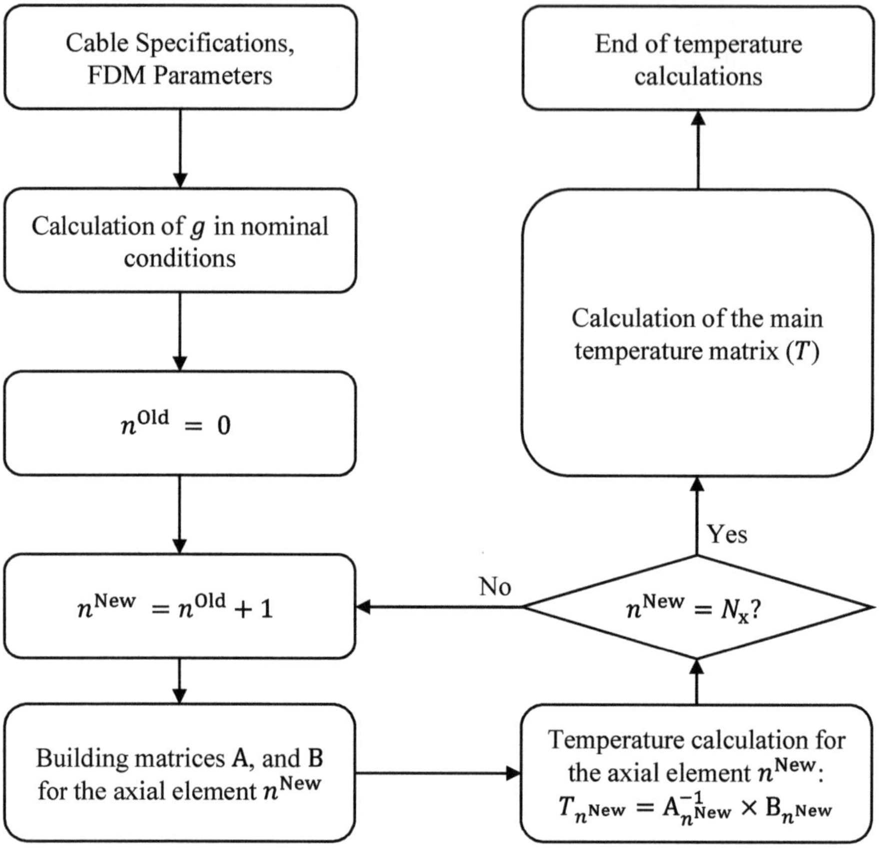

Figure 4.14: HTS cable temperature calculation algorithm in the 2-D, stationary model, where it starts from first axial element, n=1, and moves forward to the end of the cable. In this model, only the coolant temperature at the beginning of the cable is known, that is equal to LN2 temperature, 77 K.

In nominal conditions, almost all the line current passes through the HTS tapes, due to the negligible REBCO resistance. Therefore, the power generated in the other elements of solid layers would be negligible. Hence, all the arrays except arrays corresponding to the HTS tapes in the matrix $\mathrm{B_n}$ can be considered as zero in the stationary model.

4.6.2 Transient State

The heat transfer in the 2-D, transient model follows the fundamental principle given in equation (4.37) [130], [131].

$$\frac{d^2T}{dr^2} + \frac{1}{r} \cdot \frac{dT}{dr} + \frac{d^2T}{dx^2} + \frac{g}{k} = \frac{\rho \cdot c_{\mathrm{p}}}{k} \cdot \frac{dT}{dt} \tag{4.37}$$

Applying FDM to this equation is similar to the 1-D model. However, with two dimensions, a fully implicit method like the 1-D model would not be appealing due to the large sizes of the matrices; instead, a so-called ADI (Alternating Direction Implicit) numerical method is recommended to solve this equation [125]. This method divides each time-step into two half time-steps. In the first half, the temperature in the axial direction is calculated implicitly, while for the radial direction, it is done explicitly (values from the previous time step are used). In the second half, the opposite is applied.

First Half Time-step:

In the first half time-step, the temperatures in the axial direction are treated using an implicit approach. Hence, considering z representing previous time-step, and z+1, the new time-step, the label of the new half time-step, z+1/2, is assigned to the temperatures of the axial elements in the first half time-step. Conversely, for the radial direction, an explicit approach is used, meaning the temperatures are taken from the previous time-step, z. The heat transfer equation in the FDM form for a certain radial element m in the axial element n, including all boundary conditions is shown in equation (4.38).

$$
\begin{aligned}
&-(\Delta r)^2 \cdot T_{\mathrm{m,n-1}}^{\mathrm{z}+\frac{1}{2}} + \left[\psi_{\mathrm{m,n}} + 2(\Delta r)^2\right] \cdot T_{\mathrm{m,n}}^{\mathrm{z}+\frac{1}{2}} - (\Delta r)^2 \cdot T_{\mathrm{m,n+1}}^{\mathrm{z}+\frac{1}{2}} \\
&= \left[(\Delta x)^2 \cdot \left(\frac{1}{2m} - 1\right) \cdot \left(\frac{-2}{1+\eta_{\mathrm{m,m-1}}}\right)\right] \cdot T_{\mathrm{m-1,n}}^{\mathrm{z}} \\
&+ \left[\psi_{\mathrm{m,n}} - 2(\Delta x)^2 + (\Delta x)^2 \cdot \left(\frac{1}{2m} - 1\right) \cdot \left(\frac{1-\eta_{\mathrm{m,m-1}}}{1+\eta_{\mathrm{m,m-1}}}\right) + (\Delta x)^2 \cdot \left(\frac{1}{2m} + 1\right)\right. \\
&\qquad \left. \cdot \left(\frac{1-\eta_{\mathrm{m+1,m}}}{1+\eta_{\mathrm{m+1,m}}}\right)\right] \cdot T_{\mathrm{m,n}}^{\mathrm{z}} \\
&+ \left[(\Delta x)^2 \cdot \left(\frac{1}{2m} + 1\right) \cdot \left(\frac{2\eta_{\mathrm{m+1,m}}}{1+\eta_{\mathrm{m+1,m}}}\right)\right] \cdot T_{\mathrm{m+1,n}}^{\mathrm{z}} + \frac{(\Delta x \cdot \Delta r)^2}{k_{\mathrm{m}}} \cdot g_{\mathrm{m,n}}^{\mathrm{z+1}}
\end{aligned}
\tag{4.38}
$$

In this equation, the terms highlighted in red represent the radial contribution to heat transfer, while the blue terms indicate the axial contribution. The black term on the right-hand side corresponds to the power generated within the element. This equation allows for the calculation of heat transfer for all elements in the solid layers of the cable. Additionally, the parameter $\psi_{\mathrm{m,n}}$ is a self-defined parameter with equation (4.39). In this equation, the density ρ_{m} and thermal conductivity k_{m} of a specific element m are assumed to be constant along the length. However, due to potential temperature variations along the cable's length, the specific heat capacity $c_{\mathrm{p,m,n}}$ can change. Therefore, it must be calculated individually for each axial element to account for these temperature differences.

$$
\psi_{\mathrm{m,n}} = \frac{2(\Delta x \cdot \Delta r)^2 \cdot \rho_{\mathrm{m}} \cdot c_{\mathrm{p,m,n}}}{k_{\mathrm{m}} \cdot \Delta t} \tag{4.39}
$$

The additional equations for other elements can be found in the appendices Chapter 9, section A.1. The final matrix form for the heat transfer equations in this model is like equation (4.40).

$$Ax_{\mathrm{m}} \times Tx_{\mathrm{m}}^{z+\frac{1}{2}} = Qr_{\mathrm{m}}^{z} + Bx_{\mathrm{m}} = \theta x_{\mathrm{m}} \tag{4.40}$$

This equation shows the temperature calculation at each iteration of the first half time-step for a certain redial element m. In this equation, $Tx_{\mathrm{m}}^{z+\frac{1}{2}}$ is the temperature of the radial element m in x-direction (size of the matrix: $N_{\mathrm{x}} \times 1$), and the matrix Ax_{m} is the coefficients of the matrix $Tx_{\mathrm{m}}^{z+\frac{1}{2}}$ as found in the heat transfer equations (size of the matrix: $N_{\mathrm{x}} \times N_{\mathrm{x}}$). Furthermore, on the right (known) side, the matrix Qr_{m}^{z} is the radial contribution for element m (size of the matrix: $N_{\mathrm{x}} \times 1$) calculated from the values of the previous time-step, and the matrix Bx_{m} indicate the generated heat (power loss) in this element in x-direction (size of the matrix: $N_{\mathrm{x}} \times 1$). The matrix θx_{m} is a matrix to simplify the process, and it represents all the heat (power) in element m in x-direction (size of the matrix: $N_{\mathrm{x}} \times 1$). The contents of the matrices are shown in the appendices Chapter 9, section A.1. By building all the matrices, the temperature of radial element m in x-direction $Tx_{\mathrm{m}}^{z+\frac{1}{2}}$ can be calculated with equation (4.41).

$$Tx_{\mathrm{m}}^{z+\frac{1}{2}} = Ax_{\mathrm{m}}^{-1} \times \theta x_{\mathrm{m}} \tag{4.41}$$

Finally, after completing the iterative process and calculating the temperature for each radial element along the axial direction, the main temperature vector in the first half time-step $T^{z+\frac{1}{2}}$ is calculated (size of the matrix: $N_{\mathrm{r}} \times N_{\mathrm{x}}$). Each row of this matrix belongs to one radial element; therefore, the transposed form of matrix $Tx_{\mathrm{m}}^{z+\frac{1}{2}}$ is placed on the row m of the matrix $T^{z+\frac{1}{2}}$. The matrix $T^{z+\frac{1}{2}}$ is shown in equation (4.42). In this structure, the symbol "T" on the top right of each row denotes the transposed matrix.

$$T^{z+\frac{1}{2}} = \begin{bmatrix} \left[Tx_{1}^{z+\frac{1}{2}}\right]^{\mathrm{T}} \\ \vdots \\ \left[Tx_{\mathrm{m}}^{z+\frac{1}{2}}\right]^{\mathrm{T}} \\ \vdots \\ \left[Tx_{N_{\mathrm{r}}}^{z+\frac{1}{2}}\right]^{\mathrm{T}} \end{bmatrix} \tag{4.42}$$

Second Half Time-step:

In the second half time-step, the direction of calculation is reversed. For the radial r-direction, an implicit approach is used, meaning the temperatures are labeled as z+1, indicating that these radial temperatures must be calculated in this half time-step. For the axial x-direction, an explicit approach is applied, and the temperatures are labeled $z+1/2$, meaning they are taken from the results of the previous half time-step. The 2-D heat transfer equation for a certain radial element m in a certain axial element n based on equation (4.37) is as follows:

$$
\begin{aligned}
&\left[(\Delta x)^2 \cdot \left(\frac{1}{2m} - 1\right) \cdot \left(\frac{2}{\eta_{\mathrm{m,m-1}} + 1}\right)\right] \cdot T_{\mathrm{m-1,n}}^{\mathrm{z+1}} \\
&+ \left[\psi_{\mathrm{m,n}} + 2(\Delta x)^2 + (\Delta x)^2 \cdot \left(\frac{1}{2m} - 1\right) \cdot \left(\frac{\eta_{\mathrm{m,m-1}} - 1}{\eta_{\mathrm{m,m-1}} + 1}\right) + (\Delta x)^2 \cdot \left(\frac{1}{2m} + 1\right) \right. \\
&\left. \cdot \left(\frac{\eta_{\mathrm{m+1,m}} - 1}{\eta_{\mathrm{m+1,m}} + 1}\right)\right] \cdot T_{\mathrm{m,n}}^{\mathrm{z+1}} \\
&+ \left[(\Delta x)^2 \cdot \left(\frac{1}{2m} + 1\right) \cdot \left(\frac{-2\eta_{\mathrm{m+1,m}}}{\eta_{\mathrm{m+1,m}} + 1}\right)\right] \cdot T_{\mathrm{m+1,n}}^{\mathrm{z+1}} \\
&= (\Delta r)^2 \cdot T_{\mathrm{m,n-1}}^{\mathrm{z+\frac{1}{2}}} + \left[\psi_{\mathrm{m,n}} - 2(\Delta r)^2\right] \cdot T_{\mathrm{m,n}}^{\mathrm{z+\frac{1}{2}}} + (\Delta r)^2 \cdot T_{\mathrm{m,n+1}}^{\mathrm{z+\frac{1}{2}}} \\
&+ \frac{(\Delta x \cdot \Delta r)^2}{k_{\mathrm{m}}} \cdot g_{\mathrm{m,n}}^{\mathrm{z+1}}
\end{aligned}
\tag{4.43}
$$

Similar to equation (4.38) in the first half time-step, the red term represents the radial and the blue term the axial contributions. The difference between these equations is the explicit/implicit tags of the temperatures. Additionally, the black term on the right-hand side indicates the power generated in the element. Other equations can be found in the appendices Chapter 9, section A.1. Finally, the matrix form for the second half time-step can be built by having all these equations as shown in equation (4.44).

$$
Ar_{\mathrm{n}} \times Tr_{\mathrm{n}}^{\mathrm{z+1}} = Qx_{\mathrm{n}}^{\mathrm{z+\frac{1}{2}}} + Br_{\mathrm{n}} = \theta r_{\mathrm{n}} \tag{4.44}
$$

In this form, the matrix $Tr_{\mathrm{n}}^{\mathrm{z+1}}$ represent the temperature of all elements in r-direction in a certain axial element n (size of the matrix: $N_{\mathrm{r}} \times 1$), and the matrix Ar_{n} consists of the coefficients corresponding to these temperatures (size of the matrix: $N_{\mathrm{r}} \times N_{\mathrm{r}}$). Furthermore, the matrix $Qx_{\mathrm{n}}^{\mathrm{z+\frac{1}{2}}}$ indicates the contribution of temperatures in x-direction which were calculated in the previous half time-step (size of the matrix: $N_{\mathrm{r}} \times 1$), and the matrix Br_{n} is the heat (power) generated in the elements in r-direction at a certain axial element n (size of the matrix: $N_{\mathrm{r}} \times 1$). Similar to the previous half time-step, the matrix θr_{n} simplifies the calculation (Size of the matrix: $N_{\mathrm{r}} \times 1$). The contents of these matrices are detailed in the appendices Chapter 9, section A.1. By building all the matrices, the temperature of the elements in r-direction in a certain axial element n, $Tr_{\mathrm{n}}^{\mathrm{z+1}}$, can be calculated from equation (4.44) with equation (4.45).

$$Tr_{\mathrm{n}}^{\mathrm{z}+1} = Ar_{\mathrm{n}}^{-1} \times \theta r_{\mathrm{n}} \tag{4.45}$$

after completing the iterative process and calculating the temperature for each axial element, the main temperature vector $T^{\mathrm{z}+1}$ is calculated (size of the matrix: $N_{\mathrm{r}} \times N_{\mathrm{x}}$), as shown in equation (4.46). This matrix is the updated version of the matrix $T^{\mathrm{z}+\frac{1}{2}}$ which was obtained in the first half time-step. Each column of this matrix belongs to one axial element; therefore, the matrix $Tr_{\mathrm{n}}^{\mathrm{z}+1}$ is placed on the column n of the matrix T.

$$T^{\mathrm{z}+1} = \left[[Tr_1^{\mathrm{z}+1}] \quad \cdots \quad [Tr_{\mathrm{n}}^{\mathrm{z}+1}] \quad \cdots \quad Tr_{\mathrm{N_x}}^{\mathrm{z}+1}\right] \tag{4.46}$$

The overall approach to calculating temperature in the 2-D, transient model using FDM is shown in Figure 4.15. This approach simplifies all the procedure explained above to apply FDM using the ADI numerical method in the 2-D modeling of a cable.

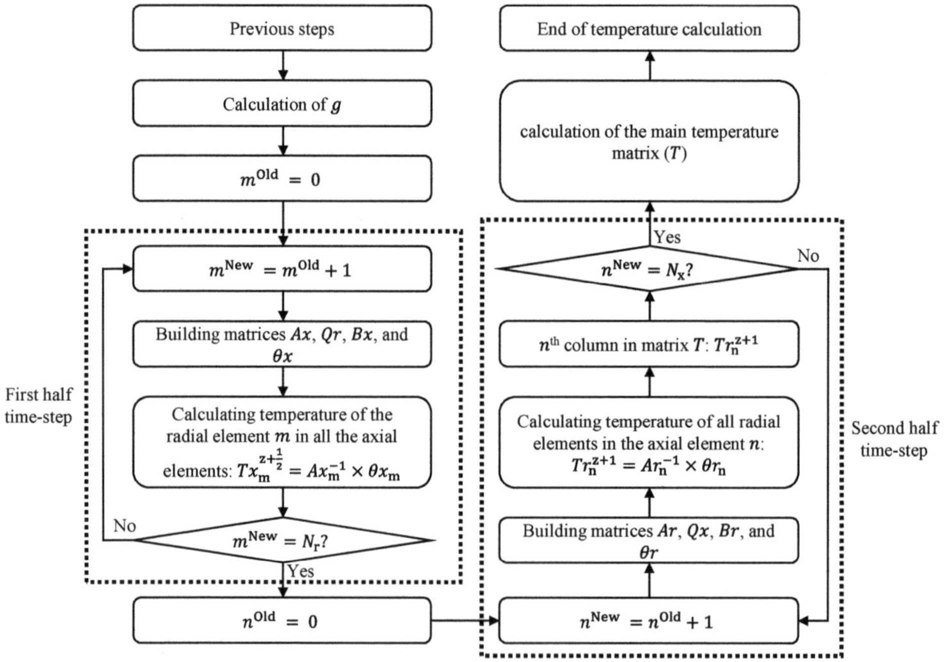

Figure 4.15: HTS cable temperature calculation algorithm in the 2-D, transient model using ADI numerical method.

The 2-D model is more complicated to implement than the 1-D and the lumped-parameter model, but it gives the most accurate results compared to reality. Especially in the case of a long-length cable, this model is vital as it can simulate the behavior of the layers, especially cooling fluid, along the cable length. Therefore, this model can be used as a base for simulating an HTS cable to verify the models with less computational efforts and complexity, lumped-parameter, and 1-D models.

4.7 Simulation

MATLAB programming is employed to simulate the behavior of the cable, using an algorithm similar to the one used for modeling the RSFCL, which was explained in Chapter 3, sub-chapter 3.6, and its flowchart presented in Figure 3.6. The key difference is that in step 2 of the algorithm, the stationary models must be simulated first to determine the initial temperature of the cable layers before proceeding with further calculations.

4.7.1 Study Cases

The electrical network considered for simulating the DC cable is the same as the study case for the RSFCL simulation, which was described in Chapter 3, sub-chapter 3.7, Figure 3.7, with the network data detailed in Table 3.2. The schematic of the study case network is depicted in Figure 4.16 where a DC pole-to-pole short circuit at load location like RSFCL simulation is studied.

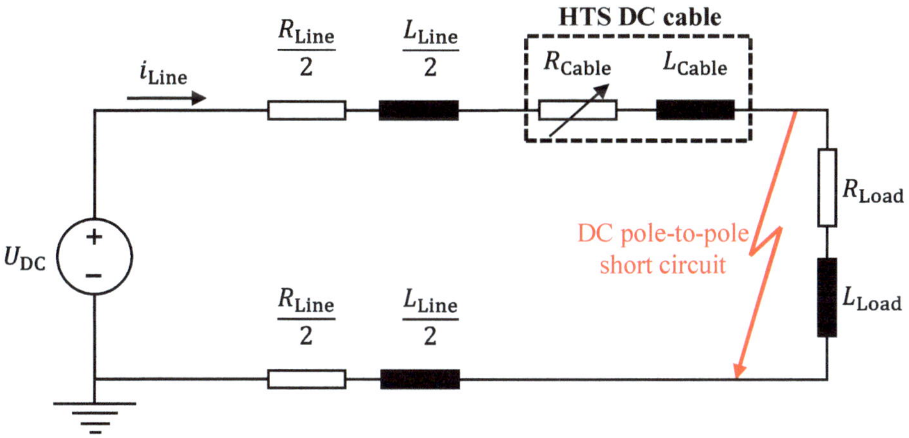

Figure 4.16: Study case network schematic to simulate HTS DC cable behavior in nominal and short circuit conditions with the components described in Chapter 3, sub-chapter 3.7, Table 3.2. In this network, the HTS DC cable equivalent circuit, which was described in Figure 4.8, is used. In this circuit, the line current, i_{Line}, is equal to the poles currents in the cable, i_{in} and i_{out}, which were described in Figure 4.9.

As the properties of the materials used in a cable change over time, the heat transfer within the cable can vary under different conditions. In the cable considered in this work, the HTS tapes are in direct contact with the copper/shield layers. While copper has very high thermal conductivity, in practice, the contact between the copper and the HTS tapes may be less effective due to factors like copper oxidation. Therefore, simulating extreme heat transfer conditions in the cable is crucial. The following cases are examined in the non-adiabatic conditions:

- **Case 1 (C1): Perfect heat conduction between HTS tapes and the core/shield copper**: This case represents the ideal scenario where optimal heat transfer occurs between the layers. To simulate this condition in the 1-D and 2-D models, no further action is needed, as the thermal conductivity of the finite-difference elements in the core/shield copper that are in direct contact with the HTS tapes is equal to the copper thermal conductivity. This case will be shown with "C1" in the simulation results.

- **Case 2 (C2): No heat conduction between HTS tapes and the core/shield copper**: In opposite to case C1, this scenario assumes no heat transfer between the layers. It is particularly relevant during transient events like short circuits, where heat generation in the HTS tapes due to fault currents is so rapid that almost no heat is transferred to the core/shield copper during the event. To simulate this condition in the 1-D and 2-D models, the thermal conductivity of the finite-difference elements in the core/shield copper that are in direct contact with the HTS tapes is set to zero. This scenario will be shown with "C2" in the simulation results.

These cases are only applicable to non-adiabatic 1-D and 2-D models. In the lumped-parameter adiabatic model, no heat conduction between layers is considered by default, so these cases are irrelevant.

4.7.2 Results

The simulation results with the developed models in different thermal conditions are explained below. The parameters illustrated in the results are described in the equivalent electrical circuit of the cable poles provided in Figure 4.9.

4.7.2.1 Lumped-parameter, adiabatic model

Figure 4.17 shows the voltage on the inner pole and the currents flowing in its different layers, including HTS tapes sub-layers.

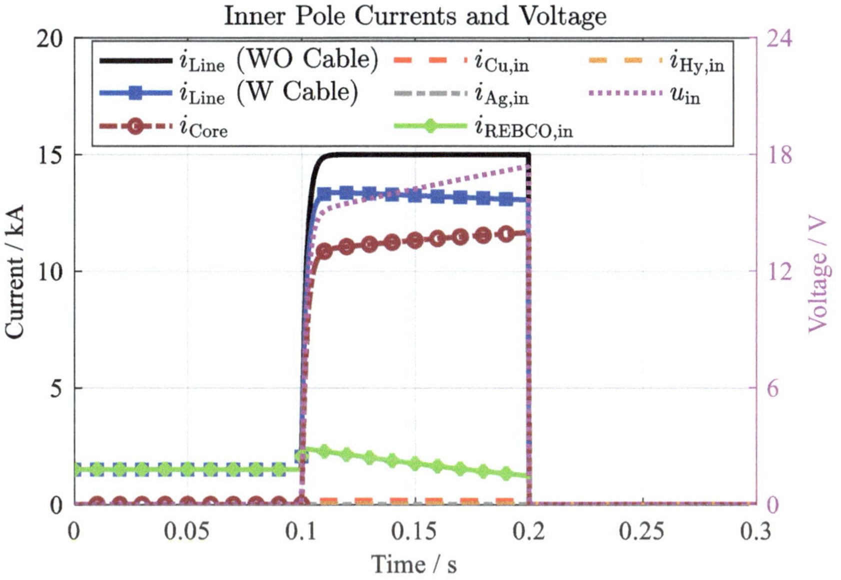

Figure 4.17: Simulation results for the study case network depicted in Figure 4.16 and the components described in Chapter 3, sub-chapter 3.7, Table 3.2 to simulate HTS DC cable behavior in nominal and short circuit conditions: the current flowing through inner pole layers, including HTS tapes, where "WO Cable" represents the network configuration without the HTS cable, while "W Cable" indicates the configuration when the HTS cable is integrated into the network. The illustrated parameters are described in Figure 4.9.

This figure illustrates that prior to the short circuit at t=0.1 s, almost all current flows through the REBCO layer of the HTS tapes. However, following the fault, the current rises, and even though it exceeds the tapes critical current, it only causes REBCO transition from flux creep to flux flow regime. This occurs due to the current shifting to the core/shield copper layers, which have relatively lower resistance than the HTS tapes. In the inner pole, most of the current flows through the core after the fault. This leads to a minor fault current limitation compared to a network with the RSFCL. The inner pole voltage, u_{in}, shows a slight increase attributed to the inner pole exceptionally low resistance, primarily due to the core copper's low resistance. This figure further depicts that other sub-layers of the HTS tapes, copper, silver, and Hastelloy, have almost no effect on the fault current due to their small cross-section area. Adding more copper to the tapes can reduce their resistance and improve heat conduction from the REBCO layer, but it may also result in a higher fault current. This approach distinguishes the tapes used in a cable from those in an RSFCL. In an RSFCL, where fault current limitation is essential, the REBCO layer is designed to quench in case of a fault, and copper is typically avoided to prevent heat dissipation. Conversely, fault limiting is usually not the primary objective of a cable. Therefore, a copper sub-layer is incorporated into the tapes to help prevent quenching of the REBCO layer.

In this regard, Figure 4.18 illustrates the tapes key parameters, including their current, temperature, and resistance. This figure also considers the parameters of all tapes in the poles. Before fault, both poles carry the same current, maintain a steady temperature at 77 K, and exhibit nearly zero resistance. After the fault occurs, their temperature and resistance begin to increase; however, due to the previously described current shift to the core/shield copper, this temperature rise remains minimal, preventing quenching. Additionally, the tapes in the inner pole show a slightly higher temperature and lower resistance due to a higher current flow, which results from the larger tape count in the outer pole (12 tapes) compared to the inner pole (8 tapes).

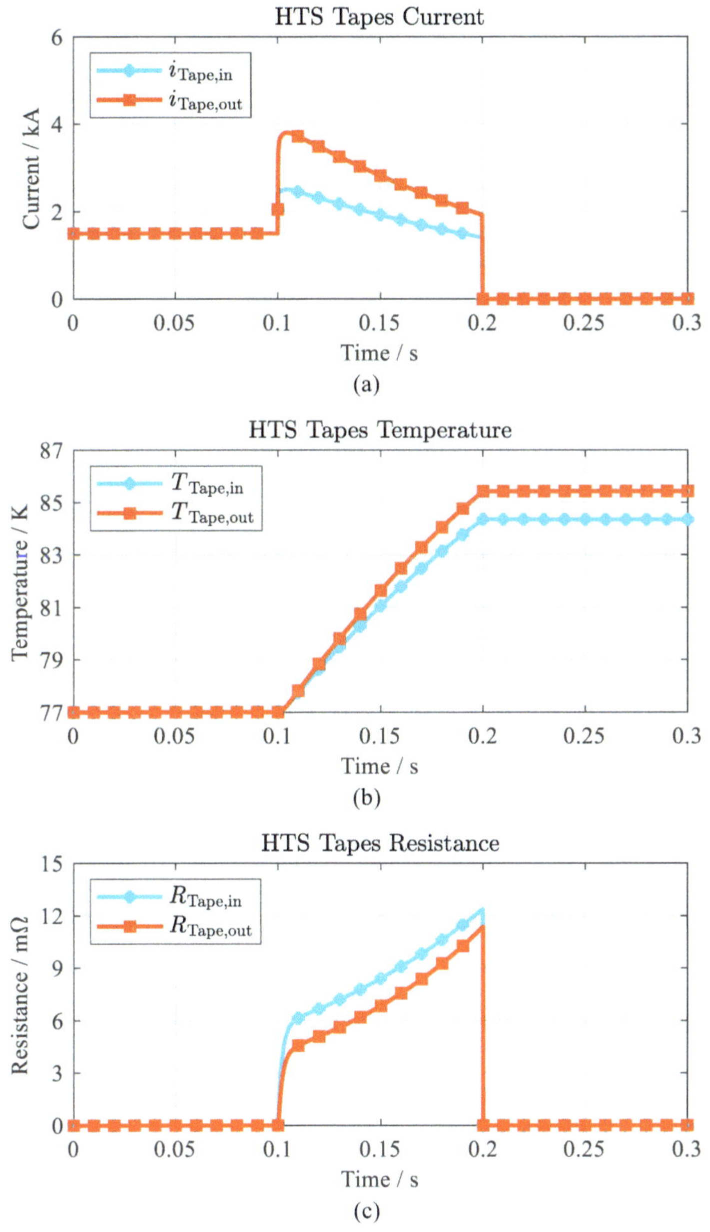

Figure 4.18: Simulation results for the study case network depicted in Figure 4.16 and the components described in Chapter 3, sub-chapter 3.7, Table 3.2 using lumped-parameter model, described in sub-chapter 0, to simulate HTS DC cable behavior in nominal and short circuit conditions: (a) HTS tapes current, where $i_{\mathrm{Tape,in}}$ and $i_{\mathrm{Tape,out}}$ represent respectively the current flowing through inner and outer poles, (b) HTS tapes temperature (both poles), where $T_{\mathrm{Tape,in}}$ and $T_{\mathrm{Tape,out}}$ represent respectively the temperature of the tapes in the inner and outer poles, (c) HTS tapes resistance (both poles), where $R_{\mathrm{Tape,in}}$ and $R_{\mathrm{Tape,out}}$ represent respectively the resistance of the tapes in the inner and outer poles.

4.7.2.2 One-dimensional, non-adiabatic model

As mentioned in the sub-chapter 4.6, to use the 1-D, transient model, first the initial temperature of the cable layers using 1-D, stationary model must be calculated. The cable temperature along its radius in the stationary mode is shown in Figure 4.19, where it can be seen that all layers have a temperature of 77 K.

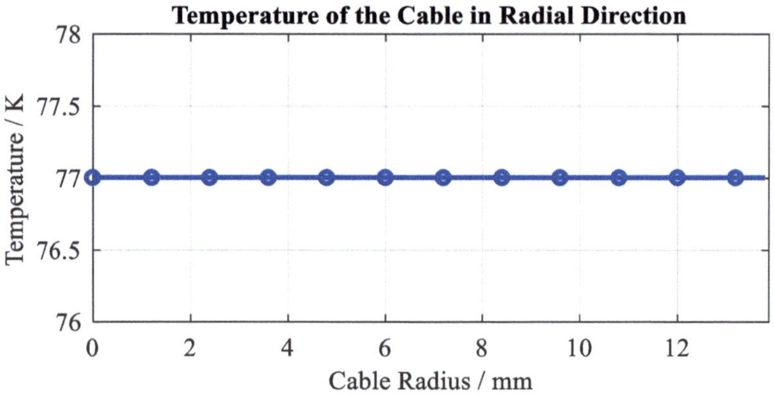

Figure 4.19: Cable temperature along its radius, resulted from 1-D, stationary model, described in sub-chapter 4.5, section 4.5.1.

The line current resulted from the simulation with the 1-D, transient model is shown in Figure 4.20, where it is compared in two cases of C1 and C2. This figure highlights a minor difference in line current between cases C1 and C2. A slightly higher limiting of the fault current in case C1 is due to the higher temperature and resistance of the conducting layers. In this case, the lack of heat transfer between tapes and core/shield copper results in a higher temperature.

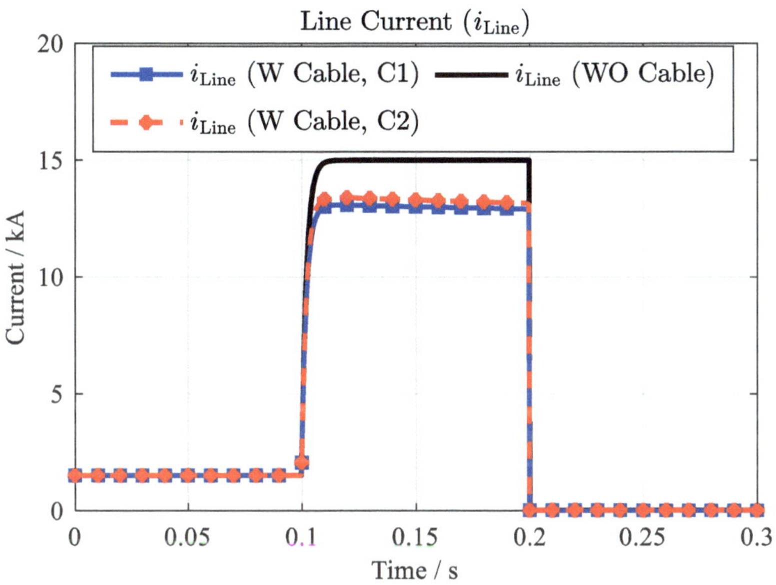

Figure 4.20: Simulation results for the study case network depicted in Figure 4.16 and the components described in Chapter 3, sub-chapter 3.7, Table 3.2 using the one-dimensional, transient model, described in sub-chapter 4.5 section 4.5.2, to simulate HTS DC cable behavior in nominal and short circuit conditions in two thermal study cases C1 and C2: line current, i_{Line}, in the network without a cable (WO Cable), and with an integrated HTS cable (W Cable).

Furthermore, the tapes temperature and resistance are depicted in Figure 4.21, where a higher temperature and resistance in case C2 can be seen because of the lack of heat transfer. This figure confirms the illustrated line current in Figure 4.20. It can also be seen that after the fault, the temperature of the tapes decreases due to the heat absorption by the coolant, LN_2. The presence of the PPLP layer, which has notably low thermal conductivity, causes heat generated in the inner pole to take longer to dissipate into the coolant compared to the outer pole. As a result, temperature changes in the HTS tapes of the outer pole occur more rapidly than in the tapes of the inner pole.

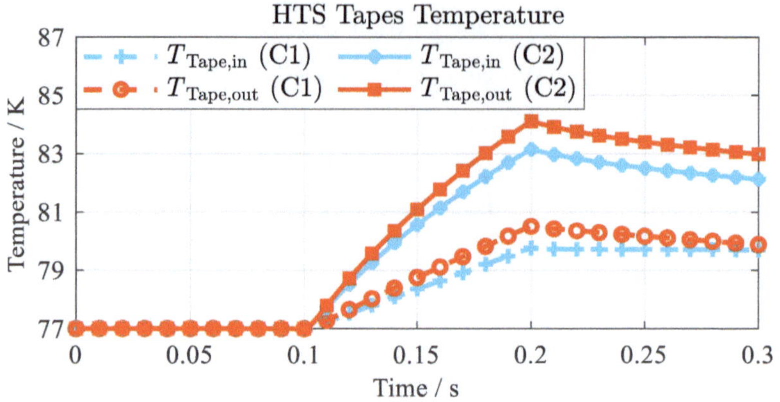

(a)

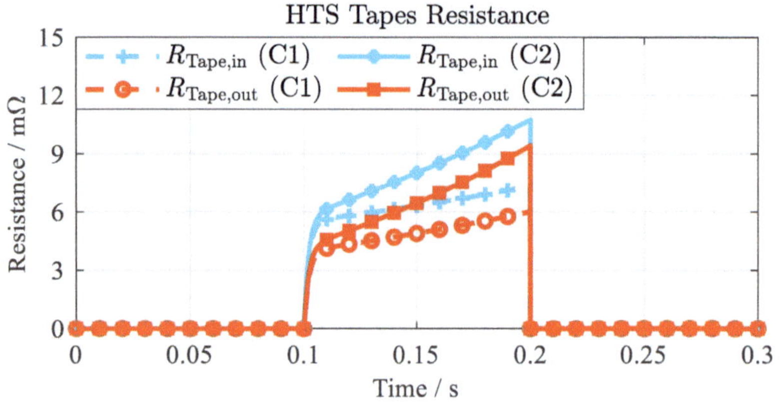

(b)

Figure 4.21: Simulation results for the study case network depicted in Figure 4.16 and the components described in Chapter 3, sub-chapter 3.7, Table 3.2 using the one-dimensional, transient model, described in sub-chapter 4.5 section 4.5.2, to simulate HTS DC cable behavior in nominal and short circuit conditions in two thermal study cases C1 and C2: (a) HTS tapes temperature (both poles), (s) HTS tapes resistance (both poles). The parameters in this figure were already presented in Figure 4.18.

An interesting feature of the 1-D, transient model is calculating the cable temperature along its radius, as illustrated in Figure 4.22. This figure shows the temperature distribution within the cable cross-section at various time intervals, highlighting heat propagation following a fault and the subsequent recovery of the HTS tapes to their superconducting state at 77 K after fault clearance. These results highlight the 1-D model capability to simulate non-adiabatic conditions and long-term heat transfer between layers (here shown at t = 40 s), a function that the lumped-parameter, adiabatic model cannot perform.

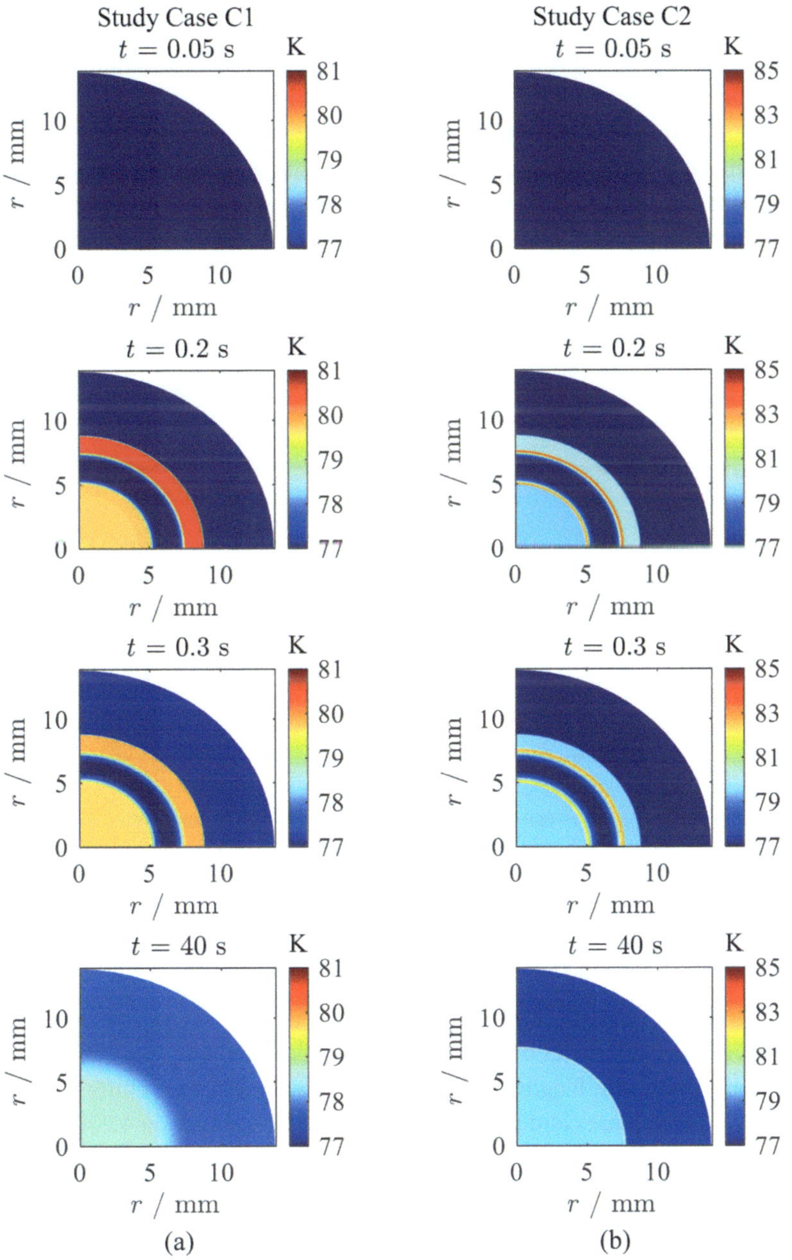

Figure 4.22: Simulation results for the study case network depicted in Figure 4.16 and the components described in Chapter 3, sub-chapter 3.7, Table 3.2 using one-dimensional, transient model, described in sub-chapter 4.5 section 4.5.2, to simulate HTS DC cable behavior in nominal and short circuit conditions in two thermal study cases at different instances before and during fault, and long-term after fault: cable temperature along its radius (cross-section) at, (a) study case 1 (C1), (b) study case 2 (C2)

4.7.2.3 Two-dimensional, non-adiabatic model

Similar to the 1-D model, the 2-D, stationary model is pivotal to determining cable's initial temperature along its radius and length, as depicted in Figure 4.23. This figure illustrates that, under nominal operation, the cable temperature at various locations is roughly 77 K. However, a temperature increase along the length of the cable is evident, primarily due to the temperature rise in the coolant. This figure is significantly compatible with the temperature rise analysis in section 4.2.4.2, and Figure 4.7, since this increase in the coolant temperature is inevitable.

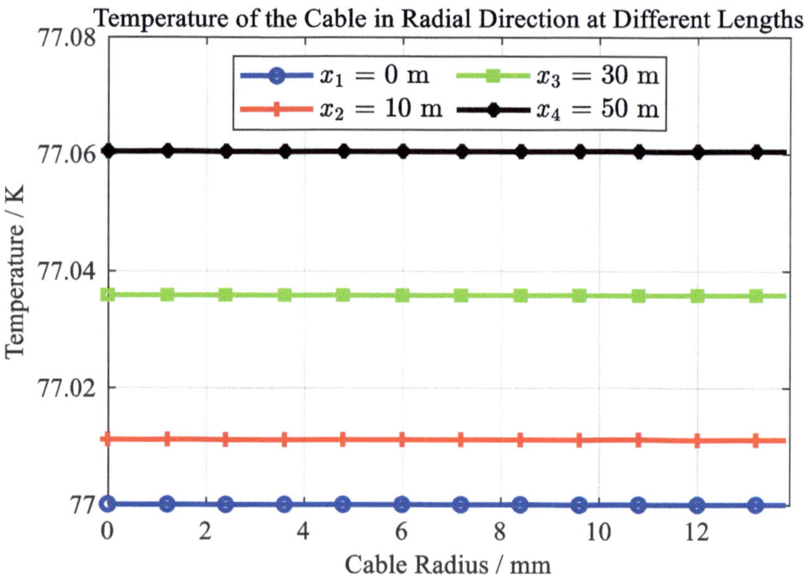

Figure 4.23: Cable temperature along its radius, resulted from 2-D, stationary model, described in sub-chapter 4.6, section 4.6.1, where x_1=0 m represents the beginning of the cable where LN2 is pumped in, and x_4=50 m indicates the end of the cable.

The advantage of the 2-D model over the 1-D model is calculating the cable temperature at different points of the cable length. This is especially essential to monitor coolant behavior since its temperature increases along the length, as explained in section 4.2.4.2 of this chapter. This is illustrated in Figure 4.24, where the increase in coolant temperature along the length is evident. After the fault, the heat generated in the tapes gradually propagates to other layers, leading to a rise in coolant temperature as it absorbs the additional heat. As the cable returns to its nominal operation following fault clearance, the coolant temperature begins to decrease (compare coolant temperature at t=10 s and t=40 s). This figure also highlights the complexity of the coolant temperature profile along the length, emphasizing the 2-D model's essential role in advanced cable analysis.

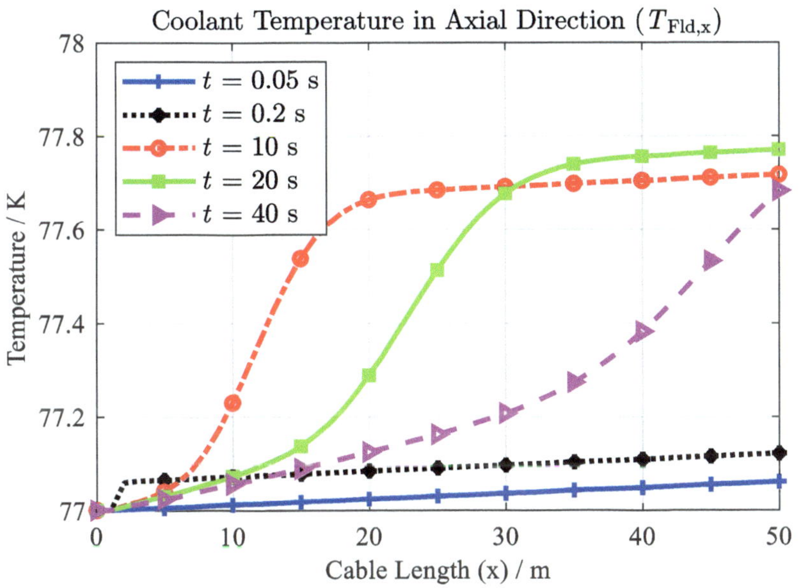

Figure 4.24: Simulation results for the study case network depicted in Figure 4.16 and the components described in Chapter 3, sub-chapter 3.7, Table 3.2 using two-dimensional, transient model, described in sub-chapter 4.6 section 4.6.2, to simulate HTS DC cable behavior in nominal and short circuit conditions: coolant (LN_2) temperature along cable length at different instances where t=0.05 s represents nominal conditions, t=0.2 s indicates 100 ms after fault, and long-term after fault clearance which is represented by t=10 s, t=20 s, and t=40 s.

Similar to the results shown in Figure 4.22 for the 1-D model, Figure 4.25 illustrates the cable temperature along its radius at t=40 s at different lengths, comparing thermal study cases C1 and C2. This figure highlights a slightly higher temperature in the tapes in case C2 due to the lack of heat dissipation. Moreover, the coolant has a higher temperature in the case of C1 due to the higher heat absorption, resulting from ideal heat transfer between layers. Another interesting observation is the PPLP role in the heat transfer between poles. Despite its significantly low thermal conductivity, it can still contribute to absorbing additional heat produced in the tapes and core/shield copper.

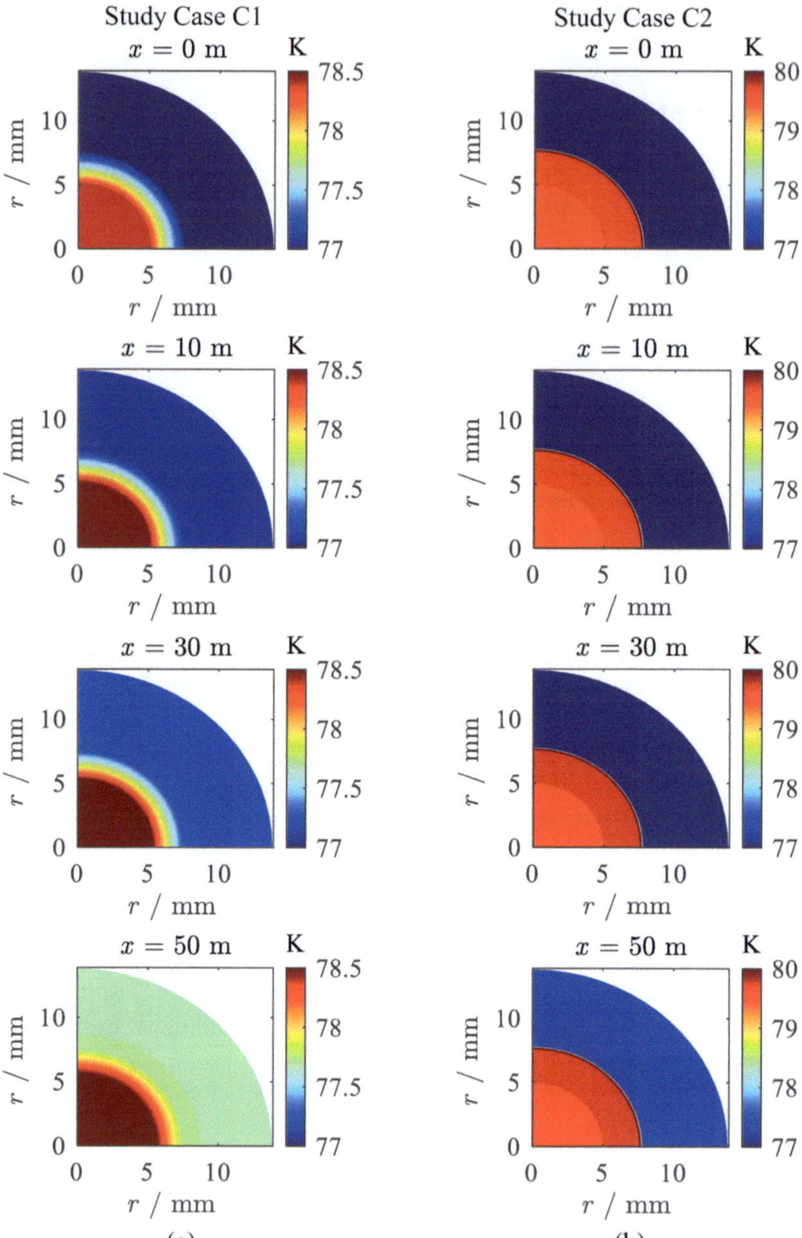

(a) (b)

Figure 4.25: Simulation results for the study case network depicted in Figure 4.16 and the components described in Chapter 3, sub-chapter 3.7, Table 3.2 using two-dimensional, transient model, described in sub-chapter 4.6 section 4.6.2, to simulate HTS DC cable behavior in nominal and short circuit conditions in two thermal study cases at t=40 s (long-term after fault clearance): cable temperature along its radius (cross-section) at, (a) study case 1 (C1), (b) study case 2 (C2).

4.7.3 Models Comparison

The models developed for the HTS cable simulation can be compared based on two main criteria:

- **Accuracy:** The models must provide accurate results that realistically simulate cable behavior. In this regard, the 2-D model is considered the most accurate as it closely adheres to physical principles. Therefore, other models' accuracy is assessed compared to the 2-D model.
- **Computational Effort:** The models should produce accurate results within a short time and with minimal computational demands.

The line current and HTS tape parameters from simulations across different models are compared to assess accuracy. Figure 4.26 illustrates the line current i_{Line} resulted from simulation with different models. This figure reveals only minor differences, validating the accuracy of the lumped-parameter and 1-D models relative to the 2-D model, considering the line current.

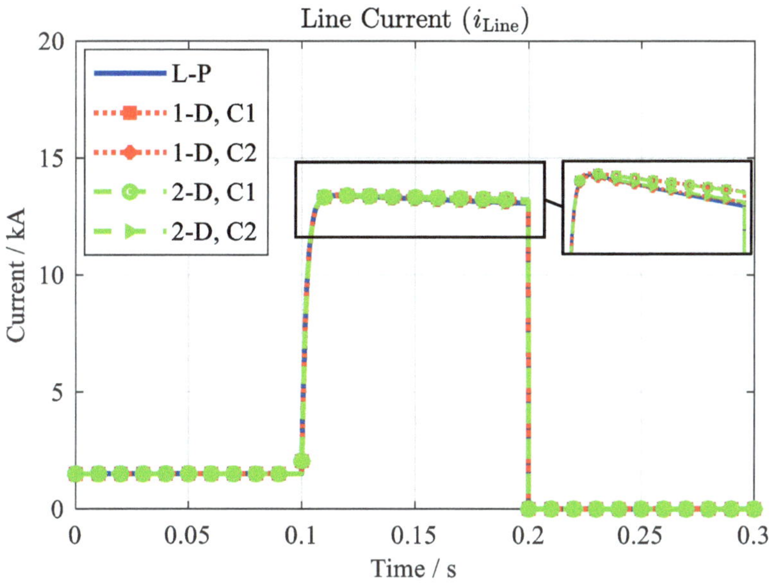

Figure 4.26: Comparison of the line current i_{Line} resulted from simulation of the study case network depicted in Figure 4.16 and the components described in Chapter 3, sub-chapter 3.7, Table 3.2 using the lumped-parameter (L-P), adiabatic model, and one-dimensional (1-D) and two-dimensional (2-D), transient models in two thermal study cases C1 and C2.

Furthermore, the temperature and resistance of the HTS tapes in the inner pole are depicted in Figure 4.27. This figure confirms that during fault conditions, the 1-D model provides results closely aligned with those of the 2-D model. Notably, under study case C2, all models yield highly similar results. As the focus is on comparing models, only the parameters of the HTS tapes in the inner pole are presented here. However, it is confirmed that the results for the HTS tapes in the outer pole follow the same principle.

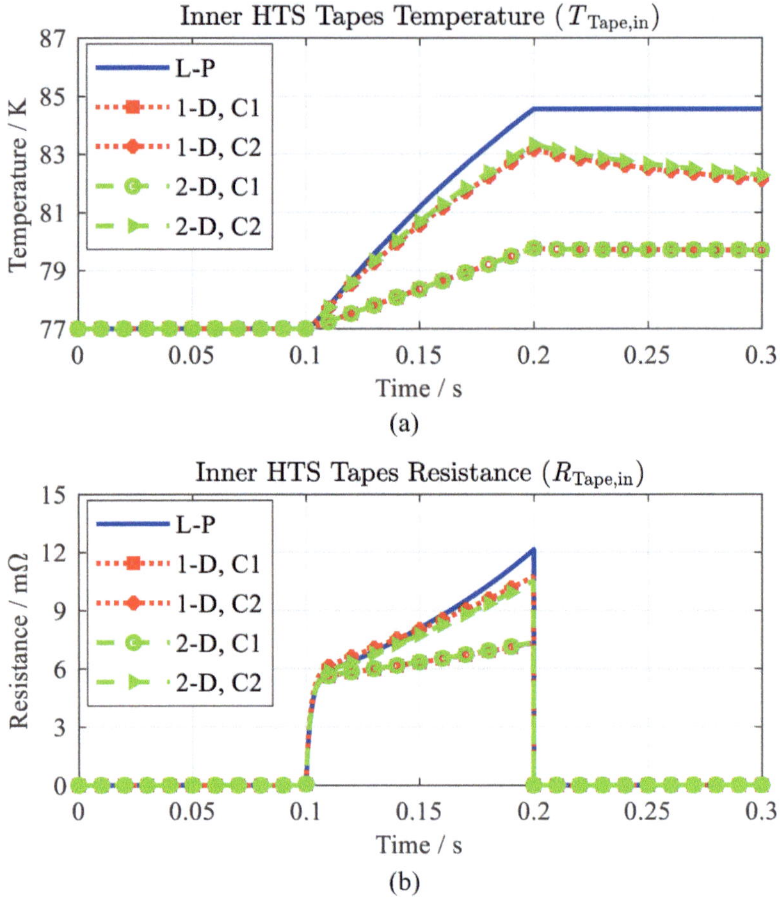

Figure 4.27: Comparison of the HTS tapes parameters resulted from simulation of the study case network depicted in Figure 4.16 and the components described in Chapter 3, sub-chapter 3.7, Table 3.2 using the lumped-parameter (L-P), adiabatic model, and one-dimensional (1-D) and two-dimensional (2-D), transient models in two thermal study cases C1 and C2: (a) Inner HTS tapes temperature $T_{\text{Tape,in}}$, (b) inner HTS tapes resistance $R_{\text{Tape,in}}$.

From a computational efforts point of view, the cable simulation for 40 s using different models on a standard computer with a 6-core 2.3 MHz CPU and 16 GB RAM takes the following times:

- Lumped-parameter, adiabatic model: instantaneous
- One-dimensional, non-adiabatic model: ~14 s
- Two-dimensional, non-adiabatic model: ~15 min

Based on these criteria, the following conclusions can be determined by comparing these models:

- **2-D Model (Most Accurate Model):** This model is crucial as it provides a basic understanding of the cable behavior under different conditions.
- **1-D Model (for Short-Length Cables and Long-Term Simulation):** For short-length cables and long-term simulations (e.g., durations lasting tens of seconds), the 1-D model yields results very close to those of the 2-D model, making it an efficient and recommended choice in these scenarios.
- **2-D Model (for Long-Length Cables):** For long cables with long-term simulations, the 2-D model becomes essential, as it is the only model that can effectively simulate coolant behavior along the cable length.
- **Lumped-Parameter Model (for Cable Simulation during a Short Circuit):** During a short circuit, the lumped-parameter model is recommended, as the parameters resulted from this model follows closely the 2-D model. Moreover, in the case of a fault, the heat generation in the HTS tapes occurs significantly fast that the thermal study case C2 (no heat conduction between tapes and core/shield copper) likely provides the most accurate representation of the cable behavior. As was seen in Figure 4.27, lumped-parameter model gives significantly similar results to the 2-D model, in the study case C2.

These models are adaptable to various scenarios. For instance, if the core and shield coppers are electrically disconnected from the voltage source, the resulting behavior could differ significantly, potentially enabling the cable to limit fault currents, a scenario that these models can simulate. Additionally, they allow for simulating cases such as a coolant pump failure in the cable system. Overall, these models provide flexibility to analyze different cases under varying conditions, depending on the specific requirements.

Following these conclusions, a SIMULINK block based on the lumped-parameter, adiabatic model is implemented to simulate the DC and AC cables behavior. Similar to the RSFCL model, these blocks are highly adjustable, allowing users to adjust parameters for various cable geometries and materials, making them adaptable for diverse cable configurations.

5 Superconducting Motor with DC/AC Inverter

5.1 Background

A lightweight and compact motor is vital for an electric aircraft due to its weight and space limitations [132]. Previous studies show that currently, an aircraft motor with a power density of at least 16 kW/kg is required, while the highest power density of a conventional permanent magnet synchronous magnet motor (PMSM) is 5.2 kW/kg [133], [134], [135]. An electric motor using superconducting technology seems to be a solution to satisfy this requirement as it allows the motor to achieve a higher power density [132], [134], [136]. This work considers a PMSM with a stator made up of superconducting tapes, as it is popular in aviation applications [137], [138], [139]. Since the objective is system modeling of the aircraft powertrain in MATLAB/SIMULINK, this motor is treated like a conventional surface-mounted PMSM with a difference of having a significantly small constant stator resistance due to the superconducting materials. Furthermore, AC losses generated by the tapes are not considered as more advanced methods like finite element method (FEM) based tools are needed that are not in the scope of this work.

The DC/AC inverter is used to drive control the motor by generating an AC voltage, which satisfies the desired inputs like speed. In the powertrain considered in this work, the inverter is placed between superconducting DC and AC cables, which operate at 77 K. Operating the inverter at the same cryogenic temperature is a sensible choice for the following reasons:

- Cryogenic power electronics provide increased power density and enhanced system performance in various applications [140], [141], [142], [143], [144], [145], [146], [147], [148]. According to [141], the experiments on an inverter for application in electric aircraft show 99% efficiency, higher than operating it at room temperature.
- The experiment in [140] shows that the resistances of the Si (Silicon) and GaN (Gallium Nitride) switches when they are ON can be significantly less at a cryogenic temperature of 77 K compared to the room temperature of 300 K. Hence, the inverter efficiency is higher at cryogenic temperatures. However, this research shows that SiC (Silicon Carbide) MOSFETs show an opposite behavior and have several times higher ON resistance at low temperatures.
- Operating the inverter at room temperature and breaking the cooling path between DC and AC cables causes unnecessary cooling losses.
- Copper lead connections between the cables and the DC/AC inverter must be larger when operated at room temperature. Therefore, running the inverter at the same cryogenic temperature as the cables would allow for smaller copper leads.

5.2 Design and Properties

The motor properties are described in Table 5.1.

Table 5.1: Superconducting motor parameters.

Parameter	Symbol	Value
Number of Pole Pairs	N_{pp}	6
Stator Resistance	R_s	10 μΩ
Stator Inductance	L_s	8.5 μH
Stator Inductance in d-axis	L_d	8.5 μH
Stator Inductance in q-axis	L_q	8.5 μH
Permanent Magnet Flux Linkage	φ_{PM}	0.2577 Wb
Maximum Speed	n_{max}	1043 rpm
Maximum (Nominal) Power	P_{nom}	456 kW
Nominal (Maximum) Torque	T_{max}	4175 N.m
Motor Inertia	J	1.19 kg.m^2
Motor Damping Coefficient	B	0.19 N.m/(rad/s)

5.3 Motor Electrical Model

The electrical equivalent circuit of the surface-mounted superconducting PMSM is depicted in Figure 5.1 [149]. In this circuit, the effect of the mutual inductance between the phases is already considered in the stator inductance L_s. A constant resistance is considered in this work, even though the stator resistance is variable with temperature and current.

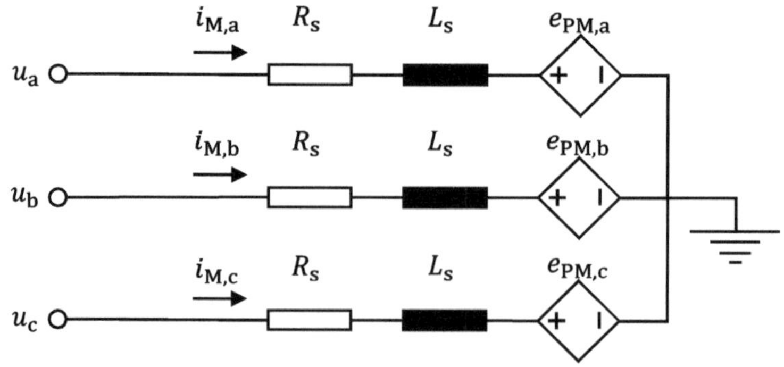

Figure 5.1: PMSM electrical equivalent circuit [149].

In this circuit, $i_{\mathrm{M,abc}}$ represent the motor three-phase currents, and $e_{\mathrm{PM,abc}}$ are the back electromotive force (EMF) induced by the permanent magnets in the three phases. Assuming a sinusoidal and harmonic-free variation of the EMF over time t, it can be calculated with equation (5.1).

$$e_{\mathrm{PM,abc}} = \varphi_{\mathrm{PM}} \cdot \omega_{\mathrm{e}} \cdot \sin(\omega_{\mathrm{e}} \cdot t + \theta_{\mathrm{abc}}) \tag{5.1}$$

In this equation, θ_{abc} are the three-phase angles given as $\theta_{\mathrm{a}} = 0$, $\theta_{\mathrm{b}} = 2\pi/3$, and $\theta_{\mathrm{b}} = 4\pi/3$. The parameter ω_{e} indicates the electrical rotational speed which is calculated from motor mechanical rotational speed ω_{m} and the number of pole pairs N_{pp} using equation (5.2).

$$\omega_{\mathrm{e}} = N_{\mathrm{pp}} \cdot \omega_{\mathrm{m}} \tag{5.2}$$

The motor speed ω_{m} itself is determined using the torque balance equation provided below. In this equation, T_{L} is the load (propeller) torque.

$$T_{\mathrm{e}} = T_{\mathrm{L}} + B \cdot \omega_{\mathrm{m}} + J \cdot \frac{d\omega_{\mathrm{m}}}{dt} \tag{5.3}$$

The torque balance equation indicates that in the case of a speed change in a motor, its torque undergoes a transient behavior, which is caused mostly by its inertia J, and a stationary (steady-state) behavior, resulted from its friction B. To design an electric aircraft powertrain, these parameters must be considered, especially in the case of a substantial speed change, the transient term could be relatively high. As the motor speed is controlled by the inverter, the rate of the speed change is decided by the drive control. Hence, the controllers must be designed accordingly to prevent transient overloading of the other components in the aircraft powertrain.

The motor three-phase voltages can then be calculated using Kirchhoff's Voltage Law as shown in Figure 5.1 with equation (5.4).

$$u_{\mathrm{abc}} = R_{\mathrm{s}} \cdot i_{\mathrm{M,abc}} + L_{\mathrm{s}} \cdot \frac{di_{\mathrm{M,abc}}}{dt} + e_{\mathrm{PM,abc}} \tag{5.4}$$

5.4 DC/AC Inverter Model

A basic inverter with a two-level topology as shown in Figure 5.2 is considered in this work [150]. In this schematic, $\mathrm{S_{abc}}$ are the inverter switches (here IGBTs are shown as an example) and their parallel diodes are shown with $\mathrm{D_{abc}}$. Additionally, the voltage and current on the DC side (generated by fuel cell) are shown with u_{DC} and i_{DC}, while the produced three-phase AC voltages and currents are depicted with u_{abc} and i_{abc}.

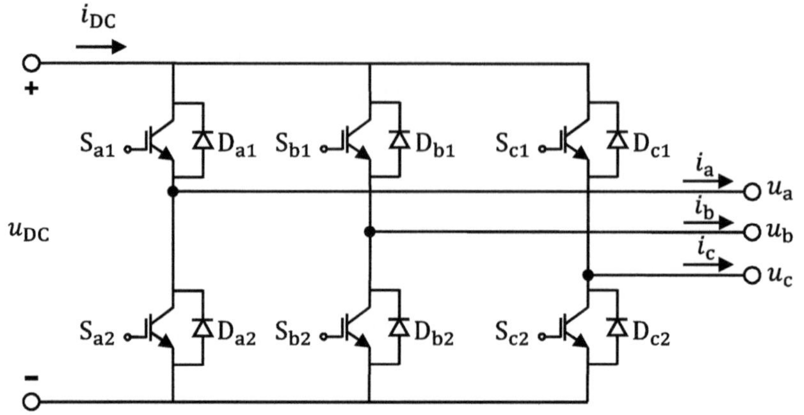

Figure 5.2: Schematic of the two-level DC/AC inverter to control the motor [150].

Since consideration of the DC/AC inverter switching effects and harmonics generated by the switches is not in the scope of this work, an average model of the inverter can be used. The equivalent average model of the DC/AC inverter is illustrated in Figure 5.3.

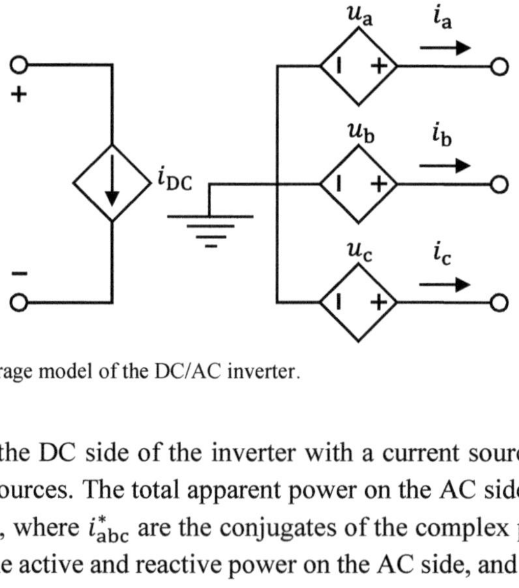

Figure 5.3: Equivalent average model of the DC/AC inverter.

This model simplifies the DC side of the inverter with a current source, and the three-phase AC side with voltage sources. The total apparent power on the AC side S_{AC} can be calculated with the equation (5.5), where i_{abc}^{*} are the conjugates of the complex phase currents. Moreover, P_{AC} and Q_{AC} are the active and reactive power on the AC side, and j is the imaginary unit.

$$S_{\mathrm{AC}} = u_{\mathrm{a}} \cdot i_{\mathrm{a}}^{*} + u_{\mathrm{b}} \cdot i_{\mathrm{b}}^{*} + u_{\mathrm{c}} \cdot i_{\mathrm{c}}^{*} = P_{\mathrm{AC}} + j \cdot Q_{\mathrm{AC}} \tag{5.5}$$

From this equation, the total active power on the AC side P_{AC} can be determined by extracting the real component of the apparent power, S_{AC}. Assuming approximately 100% efficiency in the inverter, the active power on the DC side P_{DC} is almost equal to the AC side, P_{AC}. Therefore, the DC current i_{DC} in the average model generated by the fuel cell is calculated with equation (5.6).

$$i_{\mathrm{DC}} = \frac{P_{\mathrm{DC}}}{u_{\mathrm{DC}}} = \frac{P_{\mathrm{AC}}}{u_{\mathrm{DC}}} \tag{5.6}$$

5.5 Motor Drive Control via DC/AC Inverter

Previous studies have suggested different methods for controlling a PMSM via a DC/AC inverter. These approaches fall into two main categories: scalar control, such as voltage-frequency (V/f) control, and vector control, which includes field-oriented control, voltage vector control, and direct torque control [151], [152], [153], [154], [155], [156], [157], [158]. Among all, field-oriented control is considered the most commonly known technique, as it is still the most adaptable method, offering several advantages such as reduced ripple, fast dynamic response, and high control accuracy [151], [159], [160], [161].

Figure 5.4 illustrates the field-oriented control of a PMSM. This technique consists of two loops: an outer loop for speed control and an inner loop for motor torque (current) control. Hence, while the inner loop deals with the electrical behavior of the motor, the outer loop adjusts its mechanical responses.

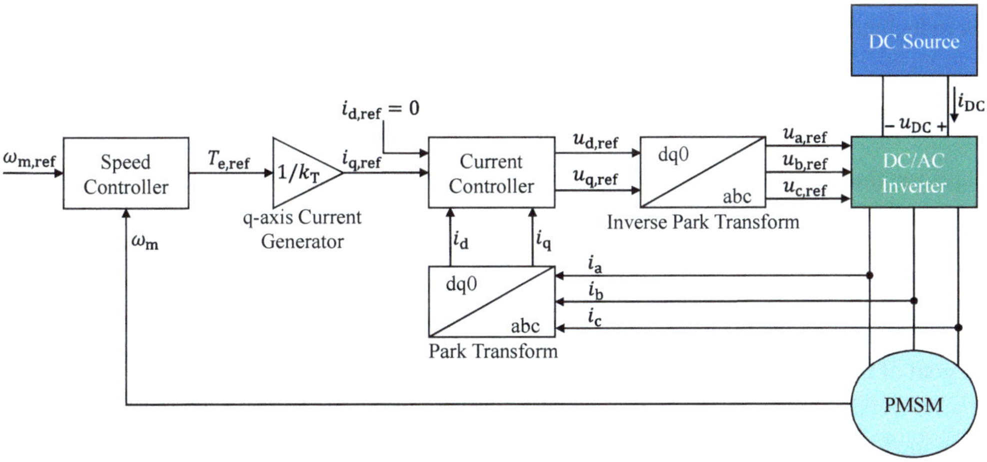

Figure 5.4: Schematic of the field-oriented control for a permanent magnet synchronous motor (PMSM) [151].

As depicted in Figure 5.4, the current control is implemented in the dq-axis reference frame, where the Park transformation (three-phase abc system to $dq0$ rotary frame) is applied to convert the inverter three-phase currents. This transformation is shown in the appendices Chapter 9, section A.1, equation (9.32) [162].

Starting from the outer loop to control the speed, the mechanical transfer function of the PMSM, $H_{\mathrm{m,PMSM}}$, as given in equation (5.7) is used.

relatively longer electrical time constant, as described in equation (5.10), which makes it comparable to the mechanical time constant. As a result, the electrical response speed becomes similar to the mechanical response speed, which is in contrast to a conventional motor, where the mechanical response is much slower. Moreover, as the torque balance equation (5.3) implies, the faster the controller is designed to respond to the speed changes, the more quasi-stationary torque is demanded by the motor due to its inertia J. Therefore, if this change by the inverter occurs significantly fast, it might end up with a substantial quasi-stationary power needed by the motor, requiring a higher amount of energy from the fuel cell stack and hydrogen. In conclusion, an appropriate controller design for such a motor in aviation applications must consider both response time and hydrogen consumption in quasi-stationary events, managing the weight and space of the hydrogen tanks.

5.6 Inverter Protection Scheme in the Short Circuit Events

As previously mentioned, a two-level topology is used for the DC/AC inverter. In this configuration, the maximum current each switch module handles in nominal conditions is equal to the maximum nominal current on its AC side. In this work, this current was given as $I_{\mathrm{AC,nom}}$ in the powertrain specifications provided in Chapter 1, Table 1.1. This nominal value is used to select suitable switches for the inverter design. However, in the event of a short circuit on the DC or AC side, the current through these switches increases. Each switch module can carry a peak forward current, which is typically 1.8 to 2 times its nominal current (for example, for a 600 V, 300 A, Infineon IGBT, this value is 600 A, which is twice its nominal current [163]). Once the current surpasses the maximum allowable level, the switches block the current flow, allowing it to pass only through the parallel diodes [150]. The equivalent circuit of the DC/AC inverter after blocking the switches is illustrated in Figure 5.7. This blocking isolates the DC and AC sides of the inverter. Since the motor rotor continues rotating, it acts as a generator, supplying current to the inverter diodes.

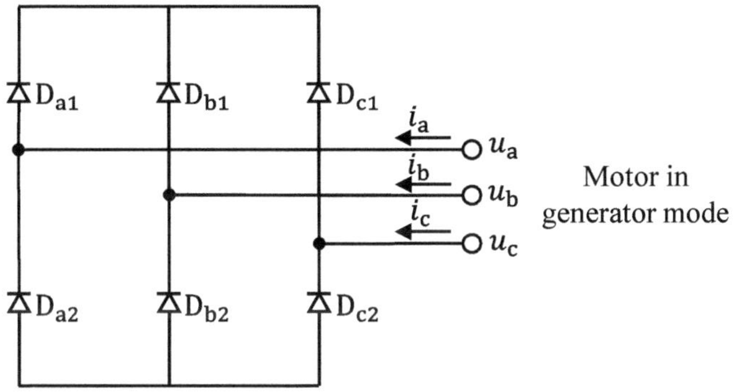

Figure 5.7: Diode-only equivalent circuit of the DC/AC inverter after blocking the switches caused by a short circuit, with electrically isolated DC and AC sides, and the motor in generator mode supplying the inverter diodes.

As depicted in Figure 5.8, an algorithm has been developed to implement this blocking behavior. Under nominal operating conditions, the motor field-oriented control regulates the three-phase voltages supplied to the AC side of the inverter's average model. However, in the event of a short circuit, these voltages are set to 0 V, and the diode-only equivalent circuit of the inverter, as shown in Figure 5.7, is activated within the network.

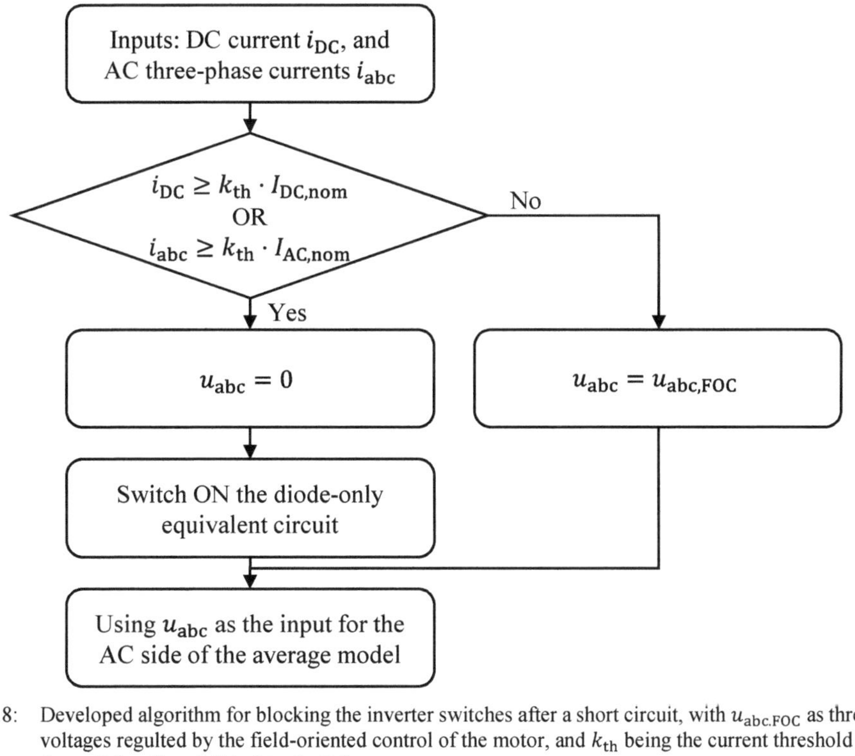

Figure 5.8: Developed algorithm for blocking the inverter switches after a short circuit, with $u_{abc.FOC}$ as three-phase voltages regulted by the field-oriented control of the motor, and k_{th} being the current threshold to block the switches (k_{th}=1.8 to 2 typically).

The final equivalent circuit of the DC/AC inverter to simulate its behavior in both nominal and short circuit conditions is demonstrated in Figure 5.9. As shown in this figure, a three-phase circuit breaker is implemented on the diode-only circuit, which is OFF in nominal conditions, and triggers ON as soon as a blocking signal is received from the inverter protection.

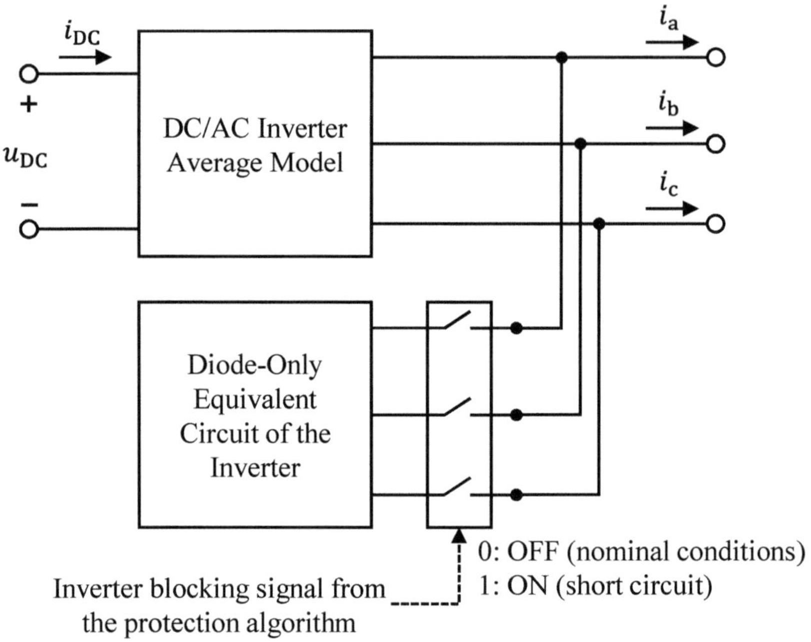

Figure 5.9: Final implementation of the DC/AC inverter circuit, with the average model supplying the motor in the nominal conditions and diode-only circuit activation after a short circuit.

Two MATLAB SIMULINK blocks were developed using the principles outlined to simulate the DC/AC inverter and motor. Like the blocks created for other components, these blocks are fully adjustable and can be adapted to various requirements. The motor model can include additional factors, but this work focuses on system-level modeling. Therefore, it targets motor performance under nominal conditions, response to speed control, and its behavior during short-circuit conditions. Considering AC loss calculation in the tapes in detail requires finite element method (FEM) based methods in tools like COMSOL Multiphysics.

5.7 Simulation

The MATLAB SIMULINK models of the DC/AC inverter and superconducting motor are used to simulate their behavior and response to speed and torque variations. In this simulation, the DC source is assumed to be a 300 V constant voltage source, and the schematic depicted in Figure 5.4 is simulated in SIMULINK. This simulation aims to validate the accurate tuning of the controllers in the field-oriented control of the superconducting motor.

5.7.1 Study Case

The motor controller behavior is simulated for both speed and torque variations. In a study case, the speed sets to reach its highest level, n_{max} as specified in Table 5.1, followed shortly by an increase in the motor torque to its maximum value, T_{max} also described in Table 5.1. These variations represent an extreme-case study scenario designed only to evaluate the tuning of the controllers. Figure 5.10 shows the motor reference speed and torque, used in this analysis.

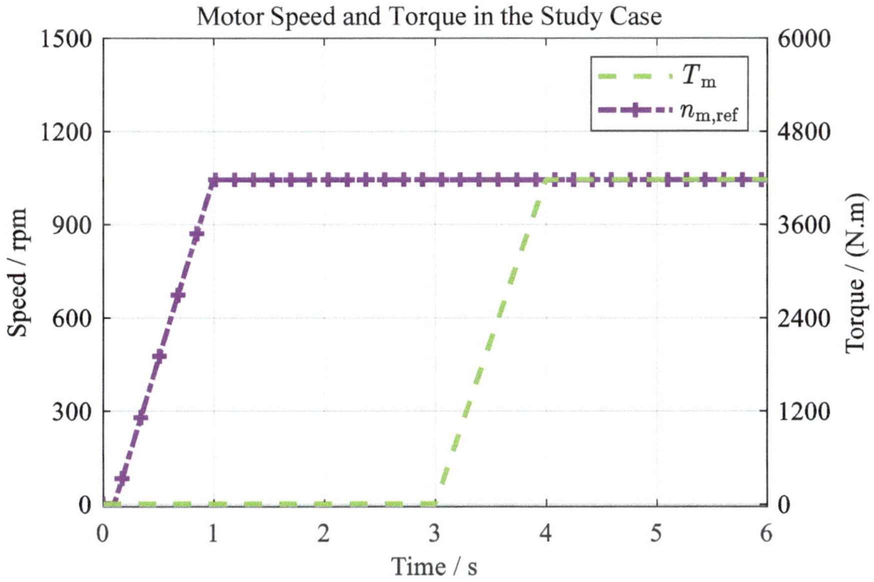

Figure 5.10: Motor reference speed $n_{m,ref}$ and toruqe T_m variations in the MATLAB SIMULINK model to simulate the field-oriented control of the superconducting motor via DC/AC inverter, with the schematic depicted in Figure 5.4.

Two control cases of C1 and C2 are analyzed in this study case, as described in Table 5.2. The parameters in case C1 were chosen to provide a significantly fast response and minimal transient error. The parameters of the current controller seem low because, as mentioned in the sub-chapter 5.5, they are tuned proportionally to the stator resistance and inductance. In the case of a superconducting PMSM, the stator resistance and inductance are low, resulting in low values for the current controller tuning. In case C2, the controller parameters are tuned ten times lower than in case C1, hence the response speed to the speed and torque variations is slower.

Table 5.2: Specifications of the motor controller parameters in two control schemes of C1 and C2.

Control Scheme	Speed Controller		Current Controller	
	$k_{p,s}$	$k_{i,s}$	$k_{p,c}$	$k_{i,c}$
C1	5950	950	0.17	0.2
C2	595	95	0.017	0.02

5.7.2 Results

Figure 5.11 illustrates the study case simulation results, where motor speed is shown. These results highlight the effectiveness of the control scheme C1 compared to the case C2. Both cases provide efficient reference speed tracking in the no-load condition, where motor torque is zero. However, by increasing the motor torque, the control scheme C1 can maintain the speed at n_{max}, the scheme C2 requires some time to regulate the speed. Therefore, the control scheme C1 is considered in this work, as it shows a better performance.

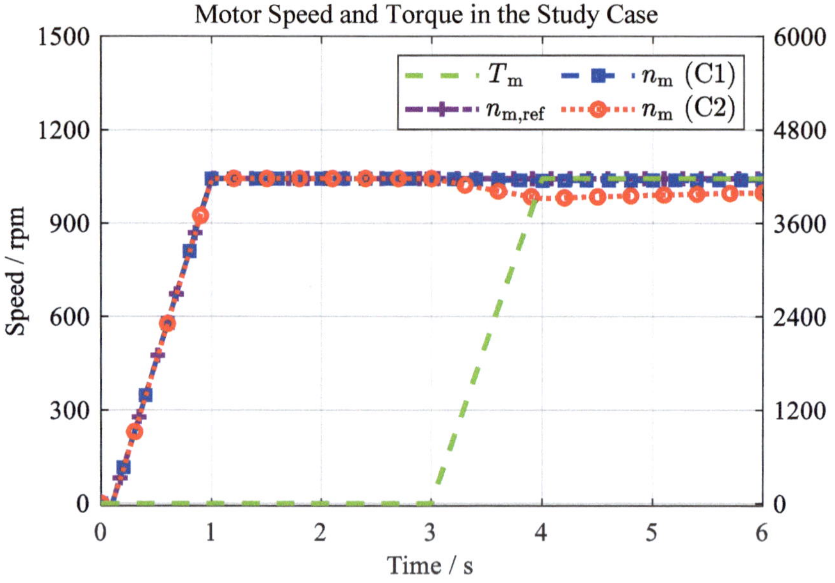

Figure 5.11: Motor speed n_m variations in the control cases C1 and C2 to the reference speed $n_{m,ref}$ and toruqe T_m changes in the MATLAB SIMULINK model to simulate the field-oriented control of the superconducting motor via DC/AC inverter, with the schematic depicted in Figure 5.4.

This analysis shows the importance of accurate parameter tuning for controlling a superconducting motor in different scenarios, ranging from no load to full-load conditions. Designing a control scheme based only on speed or torque may not be sufficient, and both parameters must be considered. Even though the target is always providing a fast response to the parameter variations, especially motor speed, the speed change rate must always be observed. As torque balance equation (5.3) implies, if the controller changes the speed significantly fast, depending on the inertia J, the quasi-stationary torque might become significantly high, resulting in a higher than expected current and power. This will substantially affect the design of the other components, causing a potentially oversizing. Moreover, since the electrical time constant of a superconducting motor is much longer than in conventional motors, and relatively close to its mechanical time constant, a compromise between current and speed controllers must always be observed, as the outputs of the speed controller is an input for the current controller. Simulating the study case for 6 s with a 6-core 2.3 MHz CPU and 16 GB RAM takes approximately one minute, highlighting the efficiency of the models in terms of computational efforts.

6 Complete Powertrain Simulation

The aircraft powertrain, with the structure shown in Chapter 1, sub-chapter 1.3, Figure 1.3, and the properties described in Table 1.1, can be simulated in MATLAB SIMULINK using the developed blocks for each component. The properties of each component are provided in their respective chapters, enabling simulation under various operating conditions. The integrated setup gives the flexibility to adjust for different power requirements, aircraft configurations, and operational flight profiles. In this work, the powertrain is simulated with reference to a typical flight profile, as well as under short circuit conditions. With these simulations, the responses to dynamic events like short circuits and speed or torque adjustments can be assessed, providing a system-level performance analysis and insight into the contribution of each single component.

6.1 Flight Profiles Description

Throughout a flight, an aircraft's propeller speed fluctuates across different phases: takeoff, cruising, and landing. An example of a realistic flight profile is shown in Figure 6.1.

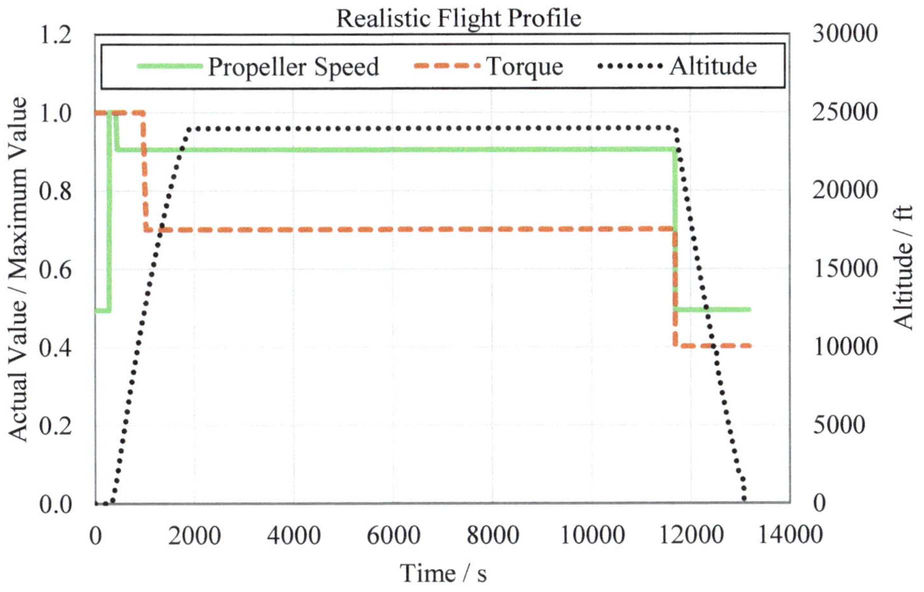

Figure 6.1: Aircraft realistic flight profile, highlighting its propeller speed, torque, and altitude from the ground, including three phases: takeoff, cruise, and landing. This flight takes almost three hours, and the aircraft undergoes different speed and torque variations [reference to internal discussions].

T_{max}, which was outlined in Chapter 5, Table 5.1. These results verify the accurate design of the fuel cell stacks based on the approach explained in Chapter 2, sub-chapter 2.2, to meet powertrain requirements.

- Figure 6.3(c) shows the variation of the fuel cell stack DC current and voltage, including its transient response to the power variations. As seen, its maximum current reaches 1.52 kA, equal to $I_{DC,nom}$ which was specified in Chapter 1, Table 1.1, while the voltage decreases to 300 V, aligned with $U_{DC,nom}$, as described in Chapter 1, Table 1.1. This result shows the necessity of a proper fuel cell stack design.
- Figure 6.3(d) depicts the motor current variations to torque changes. Here, the peak AC current reaches approximately 1.8 kA, which is in line with the motor specifications developed from requirements in Chapter 1, Table 1.1.

In this simulation, the superconducting DC/AC cables and the Resistive Superconducting Fault Current Limiter (RSFCL) operate under nominal conditions, maintaining their temperature at 77 K with LN_2 cooling. Consequently, their resistance remains negligible, making them almost lossless, allowing a highly efficient power transfer within the powertrain (greater than 97 % overall efficiency [49]).

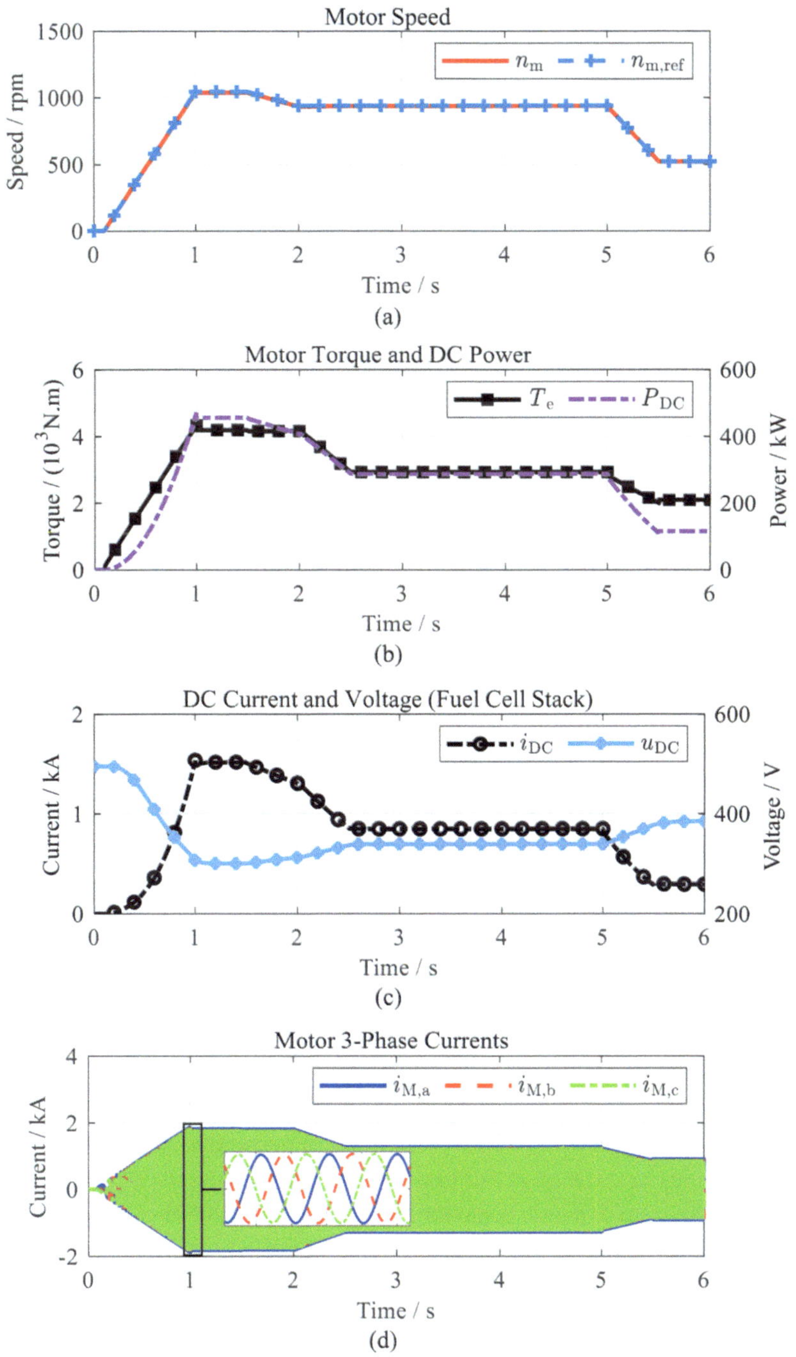

Figure 6.3: Integrated powertrain simulation results, with reference to the input speed described in the simulation flight profile, as depicted in Figure 6.2: (a) motor speed and reference speed, (b) motor torque and total DC power, (c) DC current and voltage on the output of the fuel cell stacks, (d) motor three-phase currents.

6.3 Transient Simulation (Fault Analysis)

The electric aircraft powertrain can undergo different types of short circuit. These events must be studied to analyze the components' behavior, especially with superconducting components, because their parameters can significantly change with the current increase. The integrated powertrain in SIMULINK gives the possibility of simulating any short circuit in the system, including DC pole-to-pole short circuit, three-phase, phase-to-phase, and single-phase AC short circuits, and open-circuit faults. This work analyzes the most severe cases, a DC pole-to-pole short circuit and a single-phase open-circuit fault on the AC side. The fault locations in both cases are depicted in Figure 6.4.

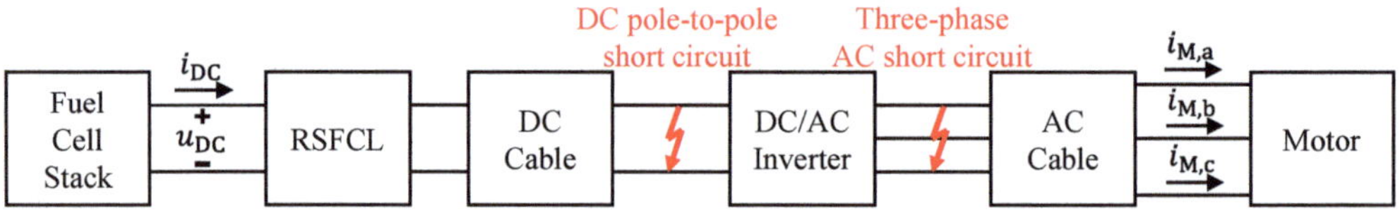

Figure 6.4: Short circuit scenarios in the aircraft powertrain, which was depicted in Chapter 1, Figure 1.1: 1) a DC pole-to-pole short circuit on the copper lead connection between DC cable and DC/AC inverter, 2) a three-phase AC short circuit on the copper lead between DC/AC inverter and AC cable. To make the worst case scenarios, in both cases, the short circuits occur when the motor rotates at its maximum speed and torque at t=1.2 s, with a fault resistance, R_f, of 1 mΩ. The duration of these short circuits is considered 100 ms. The parameters i_{DC} and u_{DC} represent the fuel cell stack current and output voltage, and $i_{M,abc}$ indicates the three-phase motor currents.

Fault Scenario 1: DC Pole-to-Pole Short Circuit

Figure 6.5 illustrates the simulation results for this short circuit, highlighting the effectiveness of the RSFCL in limiting the fault current. This figure shows that after occurring the short circuit, the current rises immediately, quenching the superconducting tapes (their temperature exceeds the critical temperature T_c) in the RSFCL. Meanwhile, the tapes in the DC and AC cables remain in a superconducting state. The figure further reveals that after fault clearance at t=1.3 s, it takes approximately 2.3 seconds for the RSFCL tapes to cool down and return to their superconducting state. This recovery time is attributed to the non-adiabatic conditions applied to the RSFCL, which allow heat dissipation to occur gradually. This delay highlights the cooling dynamics in different conditions, which were discussed in Chapter 3, section 3.5. Furthermore, the fuel cell dynamic behavior in response to a short circuit indicates that voltage drops when a fault causes current to exceed its maximum limit. However, since RSFCL limits the current, the voltage returns to the nominal operating condition shortly after. This limitation also protects the fuel cell stack from a short circuit, which might end up damaging the cells.

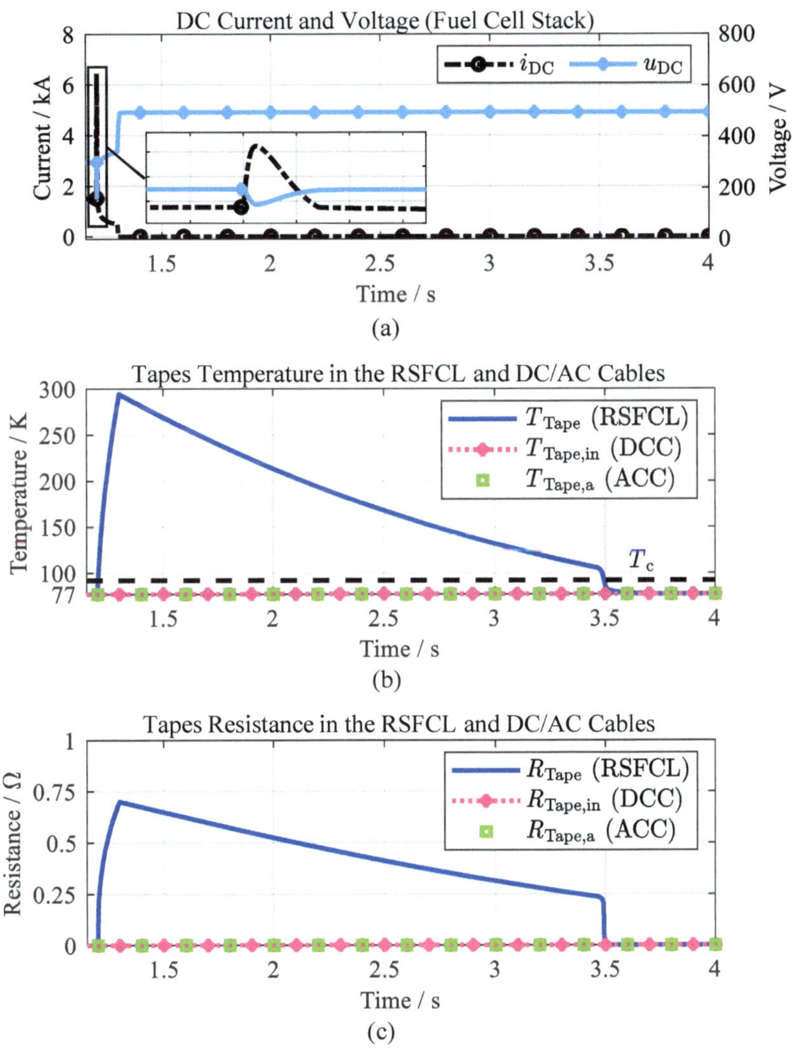

Figure 6.5: Simulation results of the DC pole-to-pole short circuit in the powertrain at the location depicted in Figure 6.4: (a) DC current i_{DC} and voltage u_{DC} at the output of the fuel cell stack, (b) tapes temperature in the RSFCL, inner pole of the DC cable (DCC) $T_{Tape,in}$, and the phase a of the AC cable (ACC) $T_{Tape,a}$, (c) total tapes resistance in the RSFLC, inner pole of the DC cable $R_{Tape,in}$, and the phase a of the AC cable $R_{Tape,a}$, where the total parallel resistance of all tapes are considered.

The RSFCL effectiveness in limiting the fault current and protecting the fuel cell can be highlighted in a comparison with a network without an RSFCL. Figure 6.6 shows the powertrain simulation in the case of a short circuit on the DC side, but in a network without an RSFCL. These results confirm that without an RSFCL, a short circuit might damage the fuel cell stack, as the DC cable cannot limit the fault current. In such a system, not only is the peak of fault current higher, but more importantly, due to not limiting it after the first peak, a significant amount of energy flows through copper leads and fuel cell electrodes. Hence, an RSFCL is recommended.

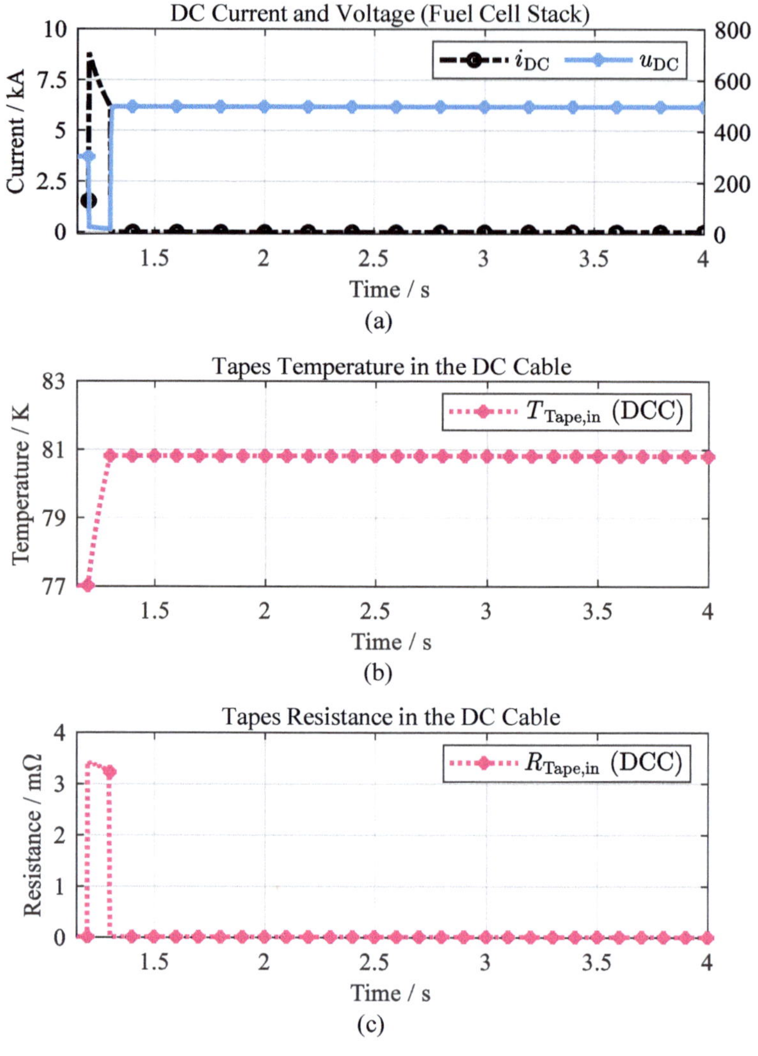

Figure 6.6: Simulation results of the DC pole-to-pole short circuit in the powertrain without an RSFCL, at the location depicted in Figure 6.4: (a) DC current i_{DC} and voltage u_{DC} at the output of the fuel cell stack, (b) tapes temperature in the inner pole of the DC cable (DCC) $T_{\mathrm{Tape,in}}$, (c) total tapes resistance in the inner pole of the DC cable $R_{\mathrm{Tape,in}}$, where the total parallel resistance of all tapes are considered.

Fault Scenario 2: Three-Phase AC Short Circuit

A three-phase short circuit on the AC side, involving the motor, is influenced significantly by the motor's inertia, since it plays a critical role in the motor's response during a short circuit.

Hence, it is also essential to consider the propeller's inertia in the analysis. The propeller's inertia depends on its design and is typically much higher than the motor's inertia. This study uses a typical value of 11.9 kg·m², ten times greater than the motor's inertia, for the propeller's inertia. This consideration ensures a more accurate representation of the powertrain dynamics during fault conditions.

The simulation results for this fault scenario are illustrated in Figure 6.7 and Figure 6.8, highlighting several key responses in the powertrain. Following the short circuit, the implemented fault protection algorithm in the DC/AC inverter blocks this component from delivering power. Figure 6.7(b) demonstrates that this blockage causes the fuel cell stack to stop delivering power, dropping its current to zero and resulting in a voltage increase to its maximum nominal level. This figure also shows that after fault, the motor rotor does not stop immediately due to inertia, and it takes a while for its speed to reach zero rpm, as observed in Figure 6.7(a). Due to this rotation, the back EMF generated by the permanent magnets in the motor still exists, leading to turning the motor into a generator feeding the fault. Consequently, a large fault current arises on the AC side, primarily due to the low stator resistance and transient short-circuit current caused by motor inductance. Figure 6.7(c) illustrates that as the motor speed decreases, the frequency of the fault current also declines. Although the back EMF includes harmonic components in practice after fault, this simulation assumes a purely sinusoidal back EMF post-fault for simplicity. It is worth mentioning that such a high current in the motor superconducting coils can be dangerous and potentially cause quenching. However, as mentioned in previous chapters, the resistance of the superconducting tapes does not remain constant as assumed in this work, and they will most likely increase due to the generated heat, leading to the fault current reduction. Hence, a more advanced model is needed to verify the electrothermal behavior of the superconducting tapes.

Furthermore, Figure 6.8 shows insightful details on the temperature and resistance changes in the superconducting tapes within the RSFCL and DC/AC cables. These figures depict that the tapes in the RSFCL and DC cable remain unaffected due to the inverter blockage. However, the AC cable directly contributes to the fault, and as observed, the tapes temperature in the phases a and b exceed critical temperature T_c due to the heat generated in them, causing the quench in the superconductor. Meanwhile, the tapes in phase c do not quench and remain in the superconducting state, primarily because of the lower fault current in the initial transient cycles. The difference in the currents flowing in each phase is caused by their different angles at the moment of the short circuit. The superconducting AC cable in the powertrain causes an increase in the fault resistance, thereby lowering fault current levels. But more importantly, it shortens the time constant for damping the energy generated by fault as fast as possible, enabling quicker recovery of the system post-fault. In this fault scenario, most of the fault current flows through the AC cable cores as their resistance is lower than the tapes resistance. This behavior limits the temperature rise in the tapes, preventing potential overheating or damage.

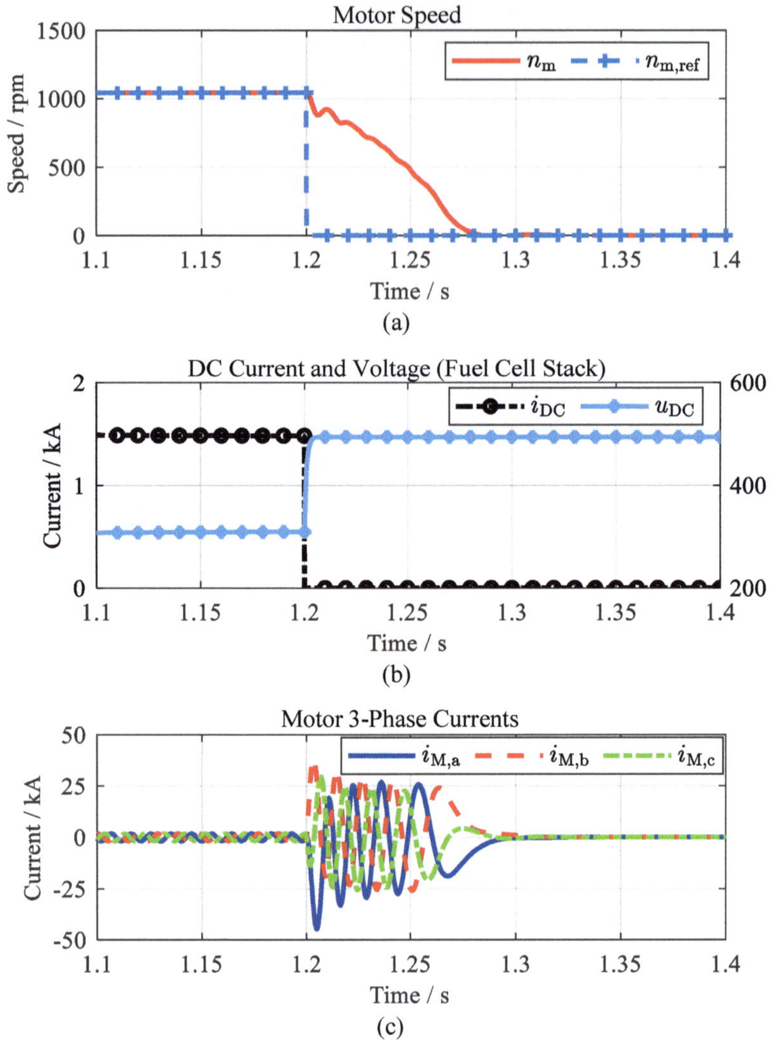

Figure 6.7: Simulation results of the three-phase AC short circuit in the powertrain at the location depicted in Figure 6.4: (a) Motor actual and reference speed, respectively denoted by n_{m} and $n_{\mathrm{m,ref}}$, (b) current i_{DC} and voltage u_{DC} at the output of the fuel cell stack, (c) motor three-phase currents.

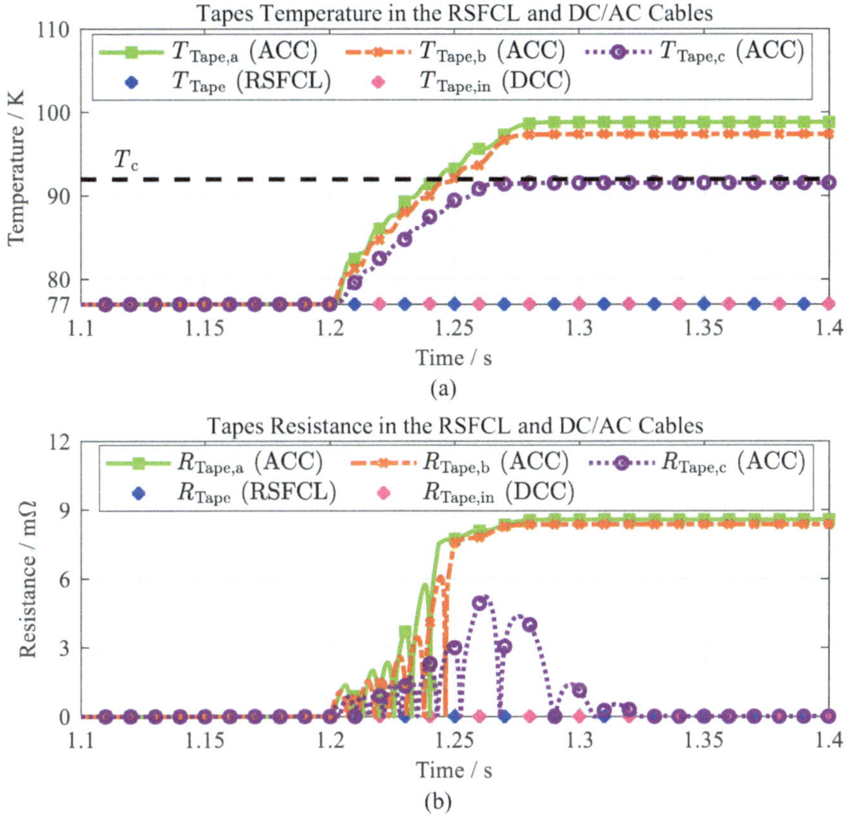

Figure 6.8: Simulation results of the three-phase AC short circuit in the powertrain at the location depicted in Figure 6.4: (a) tapes temperature in the RSFLC, inner pole of the DC cable (DCC) $T_{\mathrm{Tape,in}}$, and the three phases of the AC cable (ACC) $T_{\mathrm{Tape,abc}}$, (a) total tapes resistance in the RSFLC, inner pole of the DC cable $R_{\mathrm{Tape,in}}$, and the three phases of the AC cable (ACC) $R_{\mathrm{Tape,abc}}$, where the total parallel resistance of all tapes are considered.

6.4 Simulation Outcomes

The powertrain simulation under different conditions, as discussed in previous sections, provides insights into its performance and the contribution of each component. The following are some of the key points:

- Fuel cell stack as the power source plays a major role in the powertrain. This component has significant impacts on the system performance, not only in a nominal flight operation but also during short circuits. Therefore, a proper and careful design is needed that considers a safety margin based on the powertrain requirements.
- Using superconducting components in this powertrain increases the system efficiency, as they impose no loss in nominal conditions. This leads to lower hydrogen consumption, resulting in the advantage of requiring smaller hydrogen tanks.

- Integrating a well-designed RSFCL in the powertrain is pivotal as it can effectively limit the short circuit current, which is highly dangerous for fuel cells and DC cable. It was observed that in the case of a short circuit, the DC cable remains thermally unaffected, minimizing the risk of LN_2 evaporation within the cable. Based on the simulation results, placing an RSFCL on the AC side is also recommended, as this compact component offers substantial benefits in limiting and damping the faults, protecting the motor and AC cable.
- Tuning the motor controllers in a DC/AC inverter is more challenging than a conventional motor, as their stator resistance is much lower. This leads to a relatively longer electrical time constant, making tuning the motor response to speed variations difficult.
- A superconducting AC cable might look unnecessary in this powertrain due to its short length and minimal space savings compared to a conventional one. However, as was seen, this cable can increase the fault resistance during the short circuit, which helps the motor quickly dampen the short circuit current. Therefore, placing such a cable would be beneficial in fault management in addition to its higher efficiency and lightweight nature.
- This work treated the superconducting motor as a conventional motor with a constant substantially low stator resistance. However, this component has a multi-domain behavior, which makes it highly complicated to model in a power simulator tool like MATLAB SIMULINK. Therefore, the constant resistance assumption is used as a worst-case scenario. The simulation results indicated that the superconducting coils inside the motor, due to their negligible resistance, are exposed to a more severe fault condition compared to a conventional motor, making a superconducting motor more vulnerable to short circuits. As explained in Chapter 5, a more advanced simulation approach, such as a finite-element method (FEM), is necessary to model the superconducting motor behavior, which involves interacting electrical, thermal, and magnetic characteristics of the superconducting coils. The outputs from this work, such as the three-phase currents, could serve as input for such detailed FEM-based simulations.
- Simulation results also showed the time it takes to return to nominal operating conditions after a fault. This parameter can be used to design the protection algorithm for the DC/AC inverter.
- These simulations can give a rough estimation of the short circuit current. This estimation can be used in designing and sizing the copper leads between components.
- The integrated system in SIMULINK is significantly fast as the flight profile simulation on a computer with a 6-core 2.3 MHz CPU and 16 GB RAM takes only a few minutes.

7 Summary, Conclusions, and Outlook

7.1 Summary

This work presented the approach for modeling the powertrain of a fully electric aircraft with superconducting components in MATLAB SIMULINK. The components in such a powertrain include fuel cell stack, resistive type superconducting fault current limiter, superconducting DC and AC cables, DC/AC inverter, and superconducting motor. The work aims to simulate this powertrain in a power simulator tool like MATLAB SIMULINK, as it is pivotal to investigate system behavior under different conditions. Firstly, the simulation of all components in a standalone mode via different models and scenarios was explained. This standalone modeling allows the use of the models in wider applications and any other power systems. All the developed models were designed to provide results under stationary and transient operating conditions, allowing adaptability to various requirements. Finally, the complete powertrain simulation, including all the components in a reference flight profile and short circuit conditions, gave valuable insights into the system performance and the contribution of each component.

The fuel cell stack is the system's power supply, converting hydrogen chemical energy into electrical energy. As Proton Exchange Membrane Fuel Cells (PEMFC) offer suitable properties for aviation applications, the modeling of this fuel cell type was considered in this work. Unlike an ideal voltage source, fuel cell output voltage varies non-linearly with its current, which is called the fuel cell polarization curve. As the current required by the load increases, fuel cell voltage reduces, explained by the fuel cell stationary behavior that was simulated in this work. However, this voltage variation is not instantaneous, and the fuel cell stack undergoes a transient behavior. As was explained, most of the research articles translate this transient behavior with a resistive-capacitive electrical equivalent circuit, while in reality, a resistive-inductive behavior or a combination of both can be seen. Accordingly, a fuel cell model capable of simulating any behavior was developed, and above all, the approach to design the stack of cells based on the voltage, current, and power ratings was discussed.

An RSFCL is integrated into this powertrain to limit potential short circuits. A lumped-parameter modeling of the RSFCL, considering its electrical and thermal characteristics, was discussed. In this approach, each layer of the superconducting tapes is treated like a resistor. In an REBCO HTS tape where several layers exist, the resistance of all layers except REBCO vary linearly with temperature, while the superconductor (REBCO) resistance changes non-linearly with temperature and current. Since an HTS tape is a thin conductor, the same temperature all across the tapes was considered that varies with the heat generated by the tapes power loss. The RSFCL was simulated under two conditions: adiabatic, where coolant has no impact on the tapes, and non-adiabatic, where coolant absorbs excess generated by fault current. Simulation results indicated that heat generation in the tapes occurs quickly during the fault, and the coolant cannot contribute to heat dissipation. So, no substantial difference between adiabatic and non-adiabatic conditions was observed. After the fault, the non-adiabatic model showed an adequate contribution of the coolant in eliminating the heat and returning the tapes to the superconducting state. This recovery depends on the coolants convective heat coefficient, which varies between coolants. For example, a comparison between LN_2 and LH_2 revealed that LH_2 provides a better cooling performance, enabling a faster return of the tapes to the superconducting state.

The superconducting DC and AC cables transfer the electrical energy from the fuel cell stack to the motor. This work proposes three models to simulate the superconducting cables: lumped-parameter, adiabatic model similar to the RSFCL model, and one-dimensional and two-dimensional, finite difference method (FDM) based models. In the 1-D model, the cable temperature along its cross section is calculated, while via the 2-D model, the temperature along the length is also calculated. A bipolar DC cable can be simulated using these models under different thermal conditions. This work considered two extreme thermal conditions: perfect heat conduction between the HTS tapes and the parallel copper layers and, on the other hand, no heat conduction between them. The latter scenario seems more relevant during a short circuit, as the heat generation in the tapes occurs so quickly that no heat can be dissipated to other layers. The 2-D model provides the most accurate results, as it more precisely reflects the fundamental heat transfer principles. This model is particularly useful for a liquid coolant, showing the temperature rise along the length highlighting the risk of coolant evaporation. Therefore, with the 2-D model, the cable behavior is simulated in extreme scenarios to validate the design and eliminate the risk of coolant vaporization. Overall, this model can be used as a reference to gain a basic understanding of the cable behavior. Simulation results indicate that all models can give substantially similar results during a short circuit, while after a short circuit, the 1-D model gives similar results to the 2-D model. Therefore, the lumped-parameter model is an ideal choice for the cable simulation during a short circuit, and 1-D and 2-D models are recommended for the extended simulations. Moreover, due to the higher computational efforts and complexity of the 2-D model than the 1-D model, the 1-D model is preferred for a short-length cable, as the coolant temperature along the length does not vary substantially. The modeling of an AC cable follows the same principles as a DC cable. Thus, the SIMULINK block out of the DC/AC cable's lumped-parameter model was developed to integrate it into the powertrain.

In nominal conditions, the fuel cell stack delivers the power at a constant DC voltage, hence, a DC/AC inverter is required to transfer this power to the motor on the AC side. This work considered a superconducting PMSM treated like a conventional electric motor with a significantly small stator resistance due to the superconducting coils. The motor was modeled using an electrical circuit equivalent to a surface-mounted PMSM. This model assumed a sinusoidal harmonic-free back EMF by magnets, and the speed was calculated using the torque balance equation. The motor was controlled via DC/AC inverter using field-oriented control (FOC), the most well-known technique. Tuning the controllers for a superconducting motor is more challenging as its stator resistance is substantially low. This low resistance leads to a longer electric time constant, slowing down the electrical response to input parameters. An average model was used for the DC/AC inverter, considering no switching effects, as the objective is system-level modeling of the powertrain. Moreover, a fault protection algorithm in the inverter was developed to resemble the real behavior of the inverter in the case of a short circuit. This algorithm blocks the inverter from delivering power when a relatively higher current is detected, protecting the inverter switches from damage. As explained, a more advanced method like FEM-based tools is required to simulate a superconducting motor, but this model offers valuable insights into rough estimation of the motor behavior under different conditions. The parameters resulting from this model can be used as input for advanced simulation of the motor in FEM-based tools. In this work, a control scheme was designed to control the superconducting motor, and its effectiveness was validated via a MATLAB SIMULINK simulation.

The complete powertrain, with all the components integrated, was simulated for system-level modeling and analyzing the performance of each component. A reference flight profile drawn by a realistic flight profile was considered for nominal operating conditions. This profile included three flight phases, from takeoff to cruising and finally landing, and described the reference speed and motor torque. The reference speed was used as an input in the motor controller, and reference torque was given as load torque to the motor. First and foremost, the results indicated the importance of properly designing the components, especially the fuel cell stack as the power source. It was observed that with the increase in torque, the power and current required from the fuel cell increases, which results in the output voltage reduction. The controller's tuning was also successful, where the reference speed was followed quickly and accurately with a minor overshoot. The lossless characteristic of the superconducting components was also confirmed, as their resistance remained negligible. However, simulating various short-circuit scenarios with the integrated model is possible. The system was simulated under two severe short circuit conditions: a DC pole-to-pole short circuit and a three-phase AC short circuit. In the short circuits, the successful blocking of the inverter by the developed fault protection algorithm was observed. Simulation results showed an effective contribution by RSFCL in limiting the fault current protecting other components. This finding also supports the recommendation to integrate an RSFCL on the AC side. The fuel cell behavior during the short circuit confirmed the necessity of its accurate transient modeling. After a short circuit on the AC side, the motor turns into a generator, as its rotor keeps rotating due to propeller inertia, generating back EMF by the magnets. Since motor stator resistance is substantially low, this leads to a large short circuit current, which is dangerous for the motor coils. It was observed that even though a superconducting AC cable might not be capable of limiting this fault current, it can still help dampen the short circuit energy by adding resistance in the fault path, reducing the time constant. Moreover, this short circuit analysis is beneficial in designing and sizing the copper leads between components. The integrated system is significantly fast as the flight profile simulation on a computer with a 6-core 2.3 MHz CPU and 16 GB RAM takes only a few minutes.

7.2 Conclusions

This work successfully developed a detailed MATLAB SIMULINK powertrain model for a fuel cell-based fully electric aircraft with superconducting components. Multiple conclusions can be drawn regarding the powertrain performance and design. Some of these conclusions are provided below:

- The model enables simulation and analysis under various stationary and transient conditions by integrating potential components. The approach to model each component was explained, allowing for standalone simulation and adaptable design for broader power systems applications beyond aviation.
- A significant outcome of this work is developing a robust PEMFC model capable of capturing stationary and transient behaviors. The model was developed based on the experiments in previous research works, ensuring a realistic power source behavior for the aircraft.
- The RSFCL simulation results show that during a short circuit, the thermal considerations and cooling effect in absorbing the generated heat play a minimal role in the results. By quenching the tapes after fault, RSFCL can effectively limit the fault current, and it is recommended that an RSFCL be integrated on the AC side as well. The RSFCL model in non-adiabatic conditions is significantly helpful in determining the time after fault when coolant can absorb the additional heat. This parameter is useful in designing the protection systems of the powertrain. The simulation results indicated the significant importance of the RSFCL design in a short circuit to protect other components.
- The different models developed for the superconducting DC and AC cables showed the complex temperature variation inside the cable and the necessity of the 2-D model. However, the basic lumped-parameter model is an ideal solution during a fault, as it shows relatively similar current, temperature, and resistance behavior compared to the 2-D model. However, the developed models provide a unique opportunity to design a cable capable of limiting the fault current to eliminate the need for an RSFCL.
- This work presented the approach to developing models for the RSFCL and DC/AC cables, which are highly adaptable to the properties and type of superconductor. As an example, an RSFCL and cable with MgB_2 wires can also be simulated with this approach. Moreover, simulating these components using other coolants like LH_2 or Helium is also possible with these models. These results provide an opportunity to compare different superconductors and coolants, identifying their strengths for the application-based designs.
- Through this work, the challenges related to the drive control tuning of a superconducting motor were identified, primarily due to the low resistance of the superconducting coils of the motor. However, using the field-oriented control method with a decoupled dq-axis control, a low-error and fast-tracking of the reference parameters was explained. The tuning of the controllers is highly effective in motor behavior in quasi-

stationary events, e.g., speed and torque variations. If the drive control is designed to act significantly faster than needed, additional quasi-stationary power will be demanded by the fuel cell stack, which will result in more hydrogen being consumed. Hence, this work recommends tuning the controllers based on a compromise between the required response time and minimizing the quasi-stationary power demand. Another possibility is equipping the aircraft with onboard batteries to operate during quasi-stationary conditions. These batteries discharge during such events but can be recharged through regenerative braking when the motor slows down.

- Based on the torque balance equation, a speed change can cause a potentially high transient torque due to motor and propeller inertia. The speed variation rate is decided by the motor drive control. If the rate is too rapid, it could temporarily overload the other powertrain components. This is particularly critical for superconducting components as if they are not appropriately designed, they might undergo quenching due to overheating. In addition, according to the fuel cell stack transient behavior, this component could potentially stop delivering power as the transient current exceeds the fuel cells maximum permissible current. Therefore, the motor drive control design must be coordinated with the other components or vice versa.
- Through system simulation, the substantial importance of the fuel cell system design was observed. As fuel cell output voltage varies non-linearly with the current, stationary and quasi-stationary events must always be considered in the fuel cell stack design. Additionally, since the fuel cell system response to the load variations is not instant, the time it reaches the steady-state must be considered in the control scheme integrated into the DC/AC inverter.
- The system-level modeling of the entire powertrain, which was made possible in this work, provides the possibility to investigate the accuracy of each component's design based on various requirements. With such a model, the fuel cell design for a stable power supply in different control configurations can be examined, the copper leads between components can be appropriately sized, and the results for each component can be used as a reliable estimation for further advanced analysis and experiments.
- When a short circuit occurs on the DC side of the powertrain, the fuel cell system undergoes a severe experience that might damage its electrodes and internal components. The short circuit simulations in this work indicate that in the case of equipping no RSFCL, the peak of fault current can be limited but not significantly; however, the main advantage of the RSFCL is limiting the fault current after this first peak. When a fault is not limited afterward, substantial energy flows through the fuel cell stack and copper leads, resulting in potential damages and the requirement for unnecessary oversizing of the components.

7.3 Outlook

As this work is the first to simulate the complete superconducting powertrain of an electric aircraft, it has several perspectives. Some of them are given below:

- The primary perspective of this work is performing experiments on different components to validate the models. The real-world challenges that might have been overlooked in this work can be identified through these experiments.
- As a component involving several domains, the fuel cell stack needs to be modeled in detail with more advanced tools. Experimental analysis under different conditions would be highly beneficial in further understanding its behavior.
- In this work, the switching effects of the inverter switches and any harmonics generated by the inverter or the motor were neglected. In reality, however, these parameters might significantly impact the powertrain performance. Therefore, future works should contain these aspects as well, first in a detailed model including the power electronics behavior of the switches, then followed by converter prototypes and experiments.
- A key perspective of this work is modeling the superconducting motor. The electromagnetic behavior of the coils in the motor must be studied thoroughly as they substantially affect the superconducting tapes properties. An estimation of the current and voltages resulted from this work is significantly helpful in such analysis.
- This work considered LN_2 at 77 K as the sole coolant for the entire powertrain. As discussed in this work, different coolants have different advantages, so analyzing the feasibility of using different coolant loops for this powertrain would be beneficial. For example, LN_2 can be used to cool the cables, while helium is used as a coolant in the motor. Furthermore, the cases of losing the coolant could be interesting to analyze, as understanding how long the powertrain can operate without cooling is crucial for aircraft safety.
- Investigating the issues related to electromagnetic interference (EMI) is another interesting perspective. Since aircraft contain numerous sensors and high-frequency devices, addressing EMI is vital for operational reliability.

8 Bibliography

[1] Z. Wang and Y. Bai, "Prediction of International Aviation Carbon Emissions and Offsetting Based on ICAO Mechanisms," *2023 5th International Conference on Power and Energy Technology, ICPET 2023*, pp. 1349–1353, 2023, doi: 10.1109/ICPET59380.2023.10367631.

[2] M. Koscakova, P. Korba, I. Sekelova, P. Koscak, and D. Pastir, "Analysis of Sustainable Aviation Fuels Market," *17th International Scientific Conference on New Trends in Aviation Development, NTAD 2022 - Proceedings*, pp. 123–127, 2022, doi: 10.1109/NTAD57912.2022.10013491.

[3] M. Klöwer, M. R. Allen, D. S. Lee, S. R. Proud, L. Gallagher, and A. Skowron, "Quantifying aviation's contribution to global warming," *Environmental Research Letters*, vol. 16, no. 10, p. 104027, Nov. 2021, doi: 10.1088/1748-9326/AC286E.

[4] S. Tiwari, M. J. Pekris, and J. J. Doherty, "A review of liquid hydrogen aircraft and propulsion technologies," *Int J Hydrogen Energy*, vol. 57, pp. 1174–1196, Feb. 2024, doi: 10.1016/J.IJHYDENE.2023.12.263.

[5] B. Graver, D. Rutherford, and S. Zheng, "CO2 Emissions from Commercial Aviation 2013, 2018, and 2019," 2020.

[6] Airbus, "Global Market Forecast," Jun. 2023, Accessed: Jun. 22, 2023. [Online]. Available: https://www.airbus.com/en/products-services/commercial-aircraft/market/global-market-forecast

[7] D.-G. for M. and T. D.-G. for R. and I. European Commission, "Flightpath 2050 : Europe's vision for aviation : maintaining global leadership and serving society's needs.," *Publications Office*, 2011, doi: 10.2777/50266.

[8] M. Bradley, "Identification and Descriptions of Fuel Cell Architectures for Aircraft Applications," *2022 IEEE Transportation Electrification Conference and Expo, ITEC 2022*, pp. 1047–1050, 2022, doi: 10.1109/ITEC53557.2022.9814063.

[9] S. Blakey, L. Rye, and C. W. Wilson, "Aviation gas turbine alternative fuels: A review," *Proceedings of the Combustion Institute*, vol. 33, no. 2, pp. 2863–2885, Jan. 2011, doi: 10.1016/J.PROCI.2010.09.011.

[10] Airbus, "ZEROe - Towards the world's first hydrogen-powered commercial aircraft." Accessed: Jul. 22, 2024. [Online]. Available: https://www.airbus.com/en/innovation/energy-transition/hydrogen/zeroe

[11] N. Harrison -Communications Manager Maithili Ganesh -Project Requirements Manager, "Realising Zero-Carbon Emission Commercial Flight OUR VISION FOR ZERO-CARBON EMISSION AIR TRAVEL," 2022.

[12] W. Stautner, P. J. Ansell, and K. S. Haran, "CHEETA: An All-Electric Aircraft Takes Cryogenics and Superconductivity on Board: Combatting climate change," *IEEE Electrification Magazine*, vol. 10, no. 2, pp. 34–42, Jun. 2022, doi: 10.1109/MELE.2022.3165948.

[13] E. Cabrera and J. M. Melo de Sousa, "Use of Sustainable Fuels in Aviation—A Review," *Energies (Basel)*, vol. 15, no. 7, Apr. 2022, doi: 10.3390/EN15072440.

[14] T. Yusaf *et al.*, "Sustainable Aviation—Hydrogen Is the Future," *Sustainability (Switzerland)*, vol. 14, no. 1, Jan. 2022, doi: 10.3390/SU14010548.

[15] I. Dincer and C. Acar, "A review on potential use of hydrogen in aviation applications," *International Journal of Sustainable Aviation*, vol. 2, no. 1, p. 74, 2016, doi: 10.1504/IJSA.2016.076077.

[16] S. Sharma, S. Agarwal, and A. Jain, "Significance of hydrogen as economic and environmentally friendly fuel," *Energies (Basel)*, vol. 14, no. 21, Nov. 2021, doi: 10.3390/EN14217389.

[17] D. Višević, "Electrofuels for aviation." Accessed: Jul. 23, 2024. [Online]. Available: https://www.worldfund.vc/knowledge/electrofuels-for-aviation

[18] P. Schmidt, W. Weindorf, A. Roth, V. Batteiger, and F. Riegel, "Power-to-Liquids Potentials and Perspectives for the Future Supply of Renewable Aviation Fuel," *German Environment Agency (Umwelt Bundesamt)*, 2020, Accessed: Jul. 23, 2024. [Online]. Available: www.umweltbundesamt.de/umweltbundesamt.de/umweltbundesamt

[19] M. Ozkan *et al.*, "Forging a sustainable sky: Unveiling the pillars of aviation e-fuel production for carbon emission circularity," *iScience*, vol. 27, no. 3, p. 109154, Mar. 2024, doi: 10.1016/J.ISCI.2024.109154.

[20] A. Baroutaji, T. Wilberforce, M. Ramadan, and A. G. Olabi, "Comprehensive investigation on hydrogen and fuel cell technology in the aviation and aerospace sectors," *Renewable and Sustainable Energy Reviews*, vol. 106, pp. 31–40, May 2019, doi: 10.1016/J.RSER.2019.02.022.

[21] B. Shadidi, G. Najafi, and T. Yusaf, "A review of hydrogen as a fuel in internal combustion engines," *Energies (Basel)*, vol. 14, no. 19, Oct. 2021, doi: 10.3390/EN14196209.

[22] A. Ebrahimi, A. Rolt, S. Jafari, and J. H. Anton, "A review on liquid hydrogen fuel systems in aircraft applications for gas turbine engines," *Int J Hydrogen Energy*, vol. 91, pp. 88–105, Nov. 2024, doi: 10.1016/J.IJHYDENE.2024.10.121.

[23] P. Rompokos *et al.*, "Synergistic technology combinations for future commercial aircraft using liquid hydrogen," *J Eng Gas Turbine Power*, vol. 143, no. 7, Jul. 2021, doi: 10.1115/1.4049694.

[24] J. R. Smith and E. Mastorakos, "An energy systems model of large commercial liquid hydrogen aircraft in a low-carbon future," *Int J Hydrogen Energy*, vol. 52, pp. 633–654, Jan. 2024, doi: 10.1016/j.ijhydene.2023.04.039.

[25] B. Khandelwal, A. Karakurt, P. R. Sekaran, V. Sethi, and R. Singh, "Hydrogen powered aircraft: The future of air transport," *Progress in Aerospace Sciences*, vol. 60, pp. 45–59, 2013, doi: 10.1016/j.paerosci.2012.12.002.

[26] T. Yusaf *et al.*, "Sustainable hydrogen energy in aviation – A narrative review," *Int J Hydrogen Energy*, vol. 52, pp. 1026–1045, Jan. 2024, doi: 10.1016/J.IJHYDENE.2023.02.086.

[27] A. Joshi, R. Sharma, and B. Baral, "Comparative life cycle assessment of conventional combustion engine vehicle, battery electric vehicle and fuel cell electric vehicle in Nepal," *J Clean Prod*, vol. 379, p. 134407, Dec. 2022, doi: 10.1016/J.JCLEPRO.2022.134407.

[28] F. Qureshi *et al.*, "A State-of-The-Art Review on the Latest trends in Hydrogen production, storage, and transportation techniques," *Fuel*, vol. 340, p. 127574, May 2023, doi: 10.1016/J.FUEL.2023.127574.

[29] Q. he Luo *et al.*, "Experimental investigation of combustion characteristics and NOx emission of a turbocharged hydrogen internal combustion engine," *Int J Hydrogen Energy*, vol. 44, no. 11, pp. 5573–5584, Feb. 2019, doi: 10.1016/J.IJHYDENE.2018.08.184.

[30] M. Hepperle, "Electric Flight - Potential and Limitations," in *Energy Efficient Technologies and Concepts of Operation*, Lisbon, Portugal, Oct. 2012. Accessed: Oct. 31, 2024. [Online]. Available: https://elib.dlr.de/78726/

[31] S. Sahoo, X. Zhao, and K. Kyprianidis, "A Review of Concepts, Benefits, and Challenges for Future Electrical Propulsion-Based Aircraft," *Aerospace 2020, Vol. 7, Page 44*, vol. 7, no. 4, p. 44, Apr. 2020, doi: 10.3390/AEROSPACE7040044.

[32] Aerospace Technology Institute (ATI), "FlyZero." Accessed: Nov. 04, 2024. [Online]. Available: https://www.ati.org.uk/flyzero/

[33] J. A. Weimer, "The role of electric machines and drives in the more electric aircraft," *IEMDC 2003 - IEEE International Electric Machines and Drives Conference*, vol. 1, pp. 11–15, 2003, doi: 10.1109/IEMDC.2003.1211236.

[34] S. X. Ying, "Chapter five climate change mitigation: Aircraft technologies—Electric aircraft," 2022. doi: 10.1016/j.trc.2020.02.027.

[35] P. Pradeep and P. Wei, "Energy Optimal Speed Profile for Arrival of Tandem Tilt-Wing eVTOL Aircraft with RTA Constraint," *2018 IEEE CSAA Guidance, Navigation and Control Conference, CGNCC 2018*, Aug. 2018, doi: 10.1109/GNCC42960.2018.9018748.

[36] J. R. Smith and E. Mastorakos, "An energy systems model of large commercial liquid hydrogen aircraft in a low-carbon future," *Int J Hydrogen Energy*, vol. 52, pp. 633–654, Jan. 2024, doi: 10.1016/J.IJHYDENE.2023.04.039.

[37] H. Dae Kim, J. L. Felder, M. T. Tong, J. J. Berton, and W. J. Haller, "Turboelectric distributed propulsion benefits on the N3-X vehicle," *Aircraft Engineering and Aerospace Technology*, vol. 86, no. 6, pp. 558–561, Sep. 2014, doi: 10.1108/AEAT-04-2014-0037.

[38] C. Pornet, S. Kaiser, A. T. Isikveren, and M. Hornung, "Integrated fuel-battery hybrid for a narrow-body sized transport aircraft," *Aircraft Engineering and Aerospace Technology*, vol. 86, no. 6, pp. 568–574, Sep. 2014, doi: 10.1108/AEAT-05-2014-0062/FULL/PDF.

[39] J. Hoelzen *et al.*, "Conceptual design of operation strategies for hybrid electric aircraft," *Energies (Basel)*, vol. 11, no. 1, 2018, doi: 10.3390/EN11010217.

[40] D. Trawick, C. Perullo, M. Armstrong, D. Snyder, J. C. M. Tai, and D. N. Mavris, "Development and application of GT-HEAT to design of the Electrically Variable Engine (EVE)," *AIAA SciTech Forum - 55th AIAA Aerospace Sciences Meeting*, 2017, doi: 10.2514/6.2017-1922.

[41] J. Mangold *et al.*, "HTS machines as enabling technology for all-electric airborne vehicles," *Supercond Sci Technol*, vol. 20, no. 8, p. 748, Jun. 2007, doi: 10.1088/0953-2048/20/8/005.

[42] F. Berg, J. Palmer, P. Miller, and G. Dodds, "HTS System and Component Targets for a Distributed Aircraft Propulsion System," *IEEE Transactions on Applied Superconductivity*, vol. 27, no. 4, Jun. 2017, doi: 10.1109/TASC.2017.2652319.

[43] C. A. Luongo *et al.*, "Next generation more-electric aircraft: a potential application for hts superconductors," *IEEE Transactions on Applied Superconductivity*, vol. 19, no. 3, pp. 1055–1068, Jun. 2009, doi: 10.1109/TASC.2009.2019021.

[44] M. J. Armstrong *et al.*, "Architecture, Voltage, and Components for a Turboelectric Distributed Propulsion Electric Grid," 2015. Accessed: Oct. 31, 2024. [Online]. Available: http://www.sti.nasa.gov

[45] A. Khonya, M. Noe, W. T. B. D. Sousa, F. Berg, and M. Cooper, "Modeling Superconducting Components of the Electric Aircraft," *IEEE Transactions on Applied Superconductivity*, vol. 34, no. 3, pp. 1–5, May 2024, doi: 10.1109/TASC.2024.3350599.

[46] P. Cheetham, B. Darbha, S. Telikapalli, H. K. Chul, M. Coleman, and S. Pamidi, "Superconducting DC power distribution networks for electric aircraft," New Orleans, LA, USA: 2020 AIAA/IEEE Electric Aircraft Technologies Symposium (EATS), 2020, pp. 1–7.

[47] P. J. Masson, D. S. Soban, E. Upton, J. E. Pienkos, and C. A. Luongo, "HTS motors in aircraft propulsion: Design considerations," *IEEE Transactions on Applied Superconductivity*, vol. 15, no. 2 PART II, pp. 2218–2221, Jun. 2005, doi: 10.1109/TASC.2005.849616.

[48] A. Khonya, M. Noe, W. T. B. De Sousa, F. Berg, and M. Cooper, "Simulation Models for Superconducting Components of the Electric Aircraft," *2023 8th IEEE Workshop on the Electronic Grid, eGRID 2023*, 2023, doi: 10.1109/EGRID58358.2023.10380880.

[49] L. Ybanez *et al.*, "ASCEND: The first step towards cryogenic electric propulsion", doi: 10.1088/1757-899X/1241/1/012034.

[50] J. K. Noland, R. Mellerud, and C. Hartmann, "Next-Generation Cryo-Electric Hydrogen-Powered Aviation: A Disruptive Superconducting Propulsion System Cooled by Onboard Cryogenic Fuels," *IEEE Industrial Electronics Magazine*, vol. 16, no. 4, pp. 6–15, Dec. 2022, doi: 10.1109/MIE.2022.3174332.

[51] J. Larminie and A. Dicks, "Fuel cell systems explained: Second edition," *Fuel Cell Systems Explained: Second Edition*, pp. 1–406, Dec. 2013, doi: 10.1002/9781118878330.

[52] E. J. F. Dickinson *et al.*, "Modelling and validation of Proton exchange membrane fuel cell (PEMFC)," *IOP Conf Ser Mater Sci Eng*, vol. 290, no. 1, p. 012026, Jan. 2018, doi: 10.1088/1757-899X/290/1/012026.

[53] S. Chaudhary and Y. K. Chauhan, "Studies and performance investigations on fuel cells," *2014 International Conference on Advances in Engineering and Technology Research, ICAETR 2014*, 2014, doi: 10.1109/ICAETR.2014.7012957.

[54] S. Chaudhary and Y. K. Chauhan, "Studies and performance investigations on fuel cells," *2014 International Conference on Advances in Engineering and Technology Research, ICAETR 2014*, 2014, doi: 10.1109/ICAETR.2014.7012957.

[55] M. A. Laughton, "Fuel cells," *Engineering Science and Education Journal*, vol. 11, no. 1, pp. 7–16, 2002, doi: 10.1049/ESEJ:20020102.

[56] M. Irshad *et al.*, "A brief description of high temperature solid oxide fuel cell's operation, materials, design, fabrication technologies and performance," *Applied Sciences (Switzerland)*, vol. 6, no. 3, 2016, doi: 10.3390/APP6030075.

[57] National Energy Technology Laboratory, *Fuel Cell Handbook (Seventh Edition)*. 2004.

[58] TÜV NORD, "Hydrogen fuel cell: function & types." Accessed: Jul. 24, 2024. [Online]. Available: https://www.tuev-nord.de/en/company/energy/hydrogen/hydrogen-fuel-cell/

[59] N. Dyantyi, A. Parsons, C. Sita, and S. Pasupathi, "PEMFC for aeronautic applications: A review on the durability aspects," *Open Engineering*, vol. 7, no. 1, pp. 287–302, Feb. 2017, doi: 10.1515/ENG-2017-0035/ASSET/GRAPHIC/J_ENG-2017-0035_FIG_010.JPG.

[60] K. A. Friedrich, J. Kallo, J. Schirmer, and G. Schmitthals, "Fuel Cell Systems for Aircraft Application," *ECS Trans*, vol. 25, no. 1, pp. 193–202, Sep. 2009, doi: 10.1149/1.3210571/XML.

[61] M. Schmelcher and J. Häßy MarcSchmelcher, "Hydrogen fuel cells for aviation? A potential analysis comparing different thrust categories".

[62] D. L. Daggett, S. Eelman, and G. Kristiansson, "Fuel Cell APU for Commercial Aircraft," Jul. 2003, doi: 10.2514/6.2003-2660.

[63] J. A. Seidel, A. K. Sehra, and R. O. Colantonio, "NASA Aeropropulsion Research: Looking Forward," 2001. Accessed: Jul. 24, 2024. [Online]. Available: www.sti.nasa.gov

[64] Y. Wang, D. F. Ruiz Diaz, K. S. Chen, Z. Wang, and X. C. Adroher, "Materials, technological status, and fundamentals of PEM fuel cells – A review," *Materials Today*, vol. 32, pp. 178–203, Jan. 2020, doi: 10.1016/J.MATTOD.2019.06.005.

[65] D. M. Ali and S. K. Salman, "A comprehensive review of the fuel cells technology and hydrogen economy," *41st International Universities Power Engineering Conference, UPEC 2006, Conference Procedings*, vol. 1, pp. 98–102, 2006, doi: 10.1109/UPEC.2006.367723.

[66] X. Qingshan, W. Nianchun, K. Ichiyanagi, and K. Yukita, "PEM fuel cell modeling and parameter influences of performance evaluation," *3rd International Conference on Deregulation and Restructuring and Power Technologies, DRPT 2008*, pp. 2827–2832, 2008, doi: 10.1109/DRPT.2008.4523891.

[67] J. M. Corrêa, F. A. Farret, V. A. Popov, and M. G. Simões, "Sensitivity analysis of the modeling parameters used in simulation of proton exchange membrane fuel cells," *IEEE Transactions on Energy Conversion*, vol. 20, no. 1, pp. 211–218, Mar. 2005, doi: 10.1109/TEC.2004.842382.

[68] R. F. Mann, J. C. Amphlett, M. A. I. Hooper, H. M. Jensen, B. A. Peppley, and P. R. Roberge, "Development and application of a generalised steady-state electrochemical model for a PEM fuel cell," *J Power Sources*, vol. 86, no. 1–2, pp. 173–180, Mar. 2000, doi: 10.1016/S0378-7753(99)00484-X.

[69] W. Y. Chang, "Equivalent circuit parameters estimation for PEM fuel cell using RBF neural network and enhanced particle swarm optimization," *Math Probl Eng*, vol. 2013, 2013, doi: 10.1155/2013/672681.

[70] W. Niu, K. Song, Q. Xiao, M. Behrendt, A. Albers, and T. Zhang, "Transparency of a Geographically Distributed Test Platform for Fuel Cell Electric Vehicle Powertrain Systems Based on X-in-the-Loop Approach," *Energies 2018, Vol. 11, Page 2411*, vol. 11, no. 9, p. 2411, Sep. 2018, doi: 10.3390/EN11092411.

[71] T. C. Do *et al.*, "Energy Management Strategy of a PEM Fuel Cell Excavator with a Supercapacitor/Battery Hybrid Power Source," *Energies 2019, Vol. 12, Page 4362*, vol. 12, no. 22, p. 4362, Nov. 2019, doi: 10.3390/EN12224362.

[72] L. Xu and J. Xiao, "Modeling and simulation of PEM fuel cells based on electrochemical model," *2011 International Conference on Remote Sensing, Environment and Transportation Engineering, RSETE 2011 - Proceedings*, pp. 471–474, 2011, doi: 10.1109/RSETE.2011.5964316.

[73] F. Grumm, M. Schumann, C. Cosse, M. Plenz, A. Lücken, and D. Schulz, "Short Circuit Characteristics of PEM Fuel Cells for Grid Integration Applications," *Electronics 2020, Vol. 9, Page 602*, vol. 9, no. 4, p. 602, Apr. 2020, doi: 10.3390/ELECTRONICS9040602.

[74] F. Z. Belhaj, H. El Fadil, Z. El Idrissi, A. Intidam, M. Koundi, and F. Giri, "New Equivalent Electrical Model of a Fuel Cell and Comparative Study of Several Existing Models with Experimental Data from the PEMFC Nexa 1200 W," *Micromachines 2021, Vol. 12, Page 1047*, vol. 12, no. 9, p. 1047, Aug. 2021, doi: 10.3390/MI12091047.

[75] M. Haidoury, A. El Fatimi, H. Jbari, and M. Rachidi, "Design of a Boost DC/DC Converter Controlling a Fuel Cell System Under Proteus," *2022 International Conference on Microelectronics, ICM 2022*, pp. 46–49, 2022, doi: 10.1109/ICM56065.2022.10005332.

[76] N. Rifai, J. Sabor, and C. Alaoui, "PEM Fuel Cell Dynamic Modeling Based On Transfer Functions," *2022 2nd International Conference on Innovative Research in Applied Science, Engineering and Technology, IRASET 2022*, 2022, doi: 10.1109/IRASET52964.2022.9738273.

[77] M. Raceanu, A. Marinoiu, M. Culcer, M. Varlam, and N. Bizon, "Preventing reactant starvation of a 5 kW PEM fuel cell stack during sudden load change," *Proceedings of*

the 2014 6th International Conference on Electronics, Computers and Artificial Intelligence, ECAI 2014, pp. 55–60, 2014, doi: 10.1109/ECAI.2014.7090147.

[78] C. Yin *et al.*, "Research on Dynamic Characteristics of Fuel Cell Stack with a Hybrid Quasi-2D Equivalent Circuit Model," *IECON Proceedings (Industrial Electronics Conference)*, 2023, doi: 10.1109/IECON51785.2023.10312265.

[79] J. Zuo *et al.*, "Long-term dynamic durability test datasets for single proton exchange membrane fuel cell," *Data Brief*, vol. 35, p. 106775, Apr. 2021, doi: 10.1016/J.DIB.2021.106775.

[80] F. Z. Belhaj, H. El Fadil, Z. El Idrissi, A. Intidam, M. Koundi, and F. Giri, "New Equivalent Electrical Model of a Fuel Cell and Comparative Study of Several Existing Models with Experimental Data from the PEMFC Nexa 1200 W," *Micromachines 2021, Vol. 12, Page 1047*, vol. 12, no. 9, p. 1047, Aug. 2021, doi: 10.3390/MI12091047.

[81] O. Näckel and M. Noe, "Conceptual Design Study of an Air Coil Fault Current Limiter," *IEEE Transactions on Applied Superconductivity*, vol. 23, pp. 5602404–5602404, Apr. 2013.

[82] K. Yasuda *et al.*, "Research & development of superconducting fault current limiter in Japan," *IEEE Transactions on Applied Superconductivity*, vol. 15, no. 2, pp. 1978–1981, 2005, doi: 10.1109/TASC.2005.849349.

[83] O. Näckel and M. Noe, "Design and Test of an Air Coil Superconducting Fault Current Limiter Demonstrator," *IEEE Transactions on Applied Superconductivity*, vol. 24, no. 3, pp. 1–5, 2014, doi: 10.1109/TASC.2013.2286294.

[84] M. Noe and M. Steurer, "High-temperature superconductor fault current limiters: concepts, applications, anddevelopment status," *Supercond Sci Technol*, vol. 20, no. 3, p. R15, Jan. 2007, doi: 10.1088/0953-2048/20/3/R01.

[85] F. Mumford, "Superconducting fault current limiters," *IEE Colloquium on Fault Current Limiters - A Look at Tomorrow*, pp. 6/1-6/7, 1995.

[86] A. Morandi, "State of the art of superconducting fault current limiters and their application to the electric power system," *Physica C Supercond*, vol. 484, pp. 242–247, Jan. 2013, doi: 10.1016/J.PHYSC.2012.03.004.

[87] X. Pei, A. C. Smith, and M. Barnes, "Superconducting Fault Current Limiters for HVDC Systems," *Energy Procedia*, vol. 80, pp. 47–55, Jan. 2015, doi: 10.1016/J.EGYPRO.2015.11.405.

[88] H. Alfanan et al., "Analysing Faults and SFCL Response in Electric Aircraft," *14th European Conference on Applied Superconductivity (EUCAS 2019).*

[89] J. Xi *et al.*, "Integration of Superconducting Fault Current Limiter and DC Circuit Breaker for Electric Aircraft DC Network," *IEEE Transactions on Applied Superconductivity*, vol. 34, no. 3, pp. 1–5, May 2024, doi: 10.1109/TASC.2023.3337054.

[90] W. Song, X. Pei, J. Xi, and X. Zeng, "A Novel Helical Superconducting Fault Current Limiter for Electric Propulsion Aircraft," *IEEE Transactions on Transportation Electrification*, vol. 7, no. 1, pp. 276–286, Mar. 2021, doi: 10.1109/TTE.2020.2998417.

[91] Donglu. Shi, "High-temperature superconducting materials science and engineering : new concepts and technology," 1995.

[92] W. T. B. de Sousa, "Transient Simulations of Superconducting Fault Current Limiters," Ph.D. Dissertation, Federal University of Rio de Janeiro, Brazil, 2015.

[93] P. E. Sutherland, "Analytical model of superconducting to normal transition of bulk high Tc superconductor BSCCO-2212," *IEEE Transactions on Applied Superconductivity*, vol. 16, pp. 43–48, Mar. 2006.

[94] D. Colangelo, "Modelling of 2G HTS Coated Conductors for Fault Current Limiter Applications," École Polytechnique Fédérale de Lausanne (EPFL), Lausanne, 2013. doi: 10.5075/EPFL-THESIS-5916.

[95] J. Duron, B. Dutoit, F. Grilli, M. Decroux, L. Antognazza, and Fischer, "Computer modeling of YBCO fault current limiter strips lines in over-critical regime with temperature dependent parameters," *IEEE Transactions on Applied Superconductivity*, vol. 17, no. 2, pp. 1839–1842, Jun. 2007, doi: 10.1109/TASC.2007.898365.

[96] J. Cornejo *et al.*, "3D simulation of temperature, electric field and current density evolution in superconducting components," *Supercond Sci Technol*, vol. 16, no. 5, p. 645, Apr. 2003, doi: 10.1088/0953-2048/16/5/318.

[97] A. K. Saxena, *High-Temperature Superconductors*, 2nd ed. Heidelberg Dordrecht: Springer Science and Business Media, 2012.

[98] J. C. Butcher, *Numerical Methods for Ordinary Differential Equations*, 2nd ed. Chichester, UK: John Wiley & Sons, Ltd, 2008. doi: 10.1002/9780470753767.

[99] M. Podlaski, A. Khare, L. Vanfretti, M. Sumption, and P. Ansell, "Multi-Domain Modeling for High Temperature Superconducting Components for the CHEETA Hybrid Propulsion Power System," *AIAA Propulsion and Energy Forum, 2021*, 2021, doi: 10.2514/6.2021-3302.

[100] E. Fehlberg, "Classical Fifth-, Sixth-, Seventh-, and Eighth-Order Runge-Kutta Formulas with Stepsize Control - NASA Technical Reports Server (NTRS)." Accessed: Feb. 20, 2022. [Online]. Available: https://ntrs.nasa.gov/citations/19680027281

[101] A. Khonya, M. Noe, W. T. B. D. Sousa, F. Berg, and M. Cooper, "Modeling Superconducting Components of the Electric Aircraft," *IEEE Transactions on Applied Superconductivity*, vol. 34, no. 3, pp. 1–5, May 2024, doi: 10.1109/TASC.2024.3350599.

[102] D. W. Kim *et al.*, "Development of the 22.9-kV class HTS power cable in LG cable," *IEEE Transactions on Applied Superconductivity*, vol. 15, no. 2 PART II, pp. 1723–1726, Jun. 2005, doi: 10.1109/TASC.2005.849266.

[103] X. H. Zong, D. Wei, Y. W. Han, T. Tang, Z. Y. Zhang, and Z. G. Yu, "Development of 35kV 2000A CD HTS cable demonstration project," *2015 IEEE International Conference on Applied Superconductivity and Electromagnetic Devices, ASEMD 2015 - Proceedings*, pp. 593–595, Apr. 2016, doi: 10.1109/ASEMD.2015.7453717.

[104] V. Pothavajhala, L. Graber, C. H. Kim, and S. Pamidi, "Experimental and model based studies on current distribution in superconducting DC Cables," *IEEE Transactions on Applied Superconductivity*, vol. 24, no. 3, 2014, doi: 10.1109/TASC.2013.2282568.

[105] M. Yazdani-Asrami, S. Seyyedbarzegar, A. Sadeghi, W. T. B. De Sousa, and D. Kottonau, "High temperature superconducting cables and their performance against short circuit faults: current development, challenges, solutions, and future trends," *Supercond Sci Technol*, vol. 35, no. 8, p. 083002, Jul. 2022, doi: 10.1088/1361-6668/AC7AE2.

[106] V. E. Sytnikov *et al.*, "Status of HTS cable link project for St. Petersburg Grid," *IEEE Transactions on Applied Superconductivity*, vol. 25, no. 3, Jun. 2015, doi: 10.1109/TASC.2014.2373814.

[107] S. Mukoyama *et al.*, "Manufacturing and installation of the world's longest HTS cable in the super-ACE project," *IEEE Trans. Appl. Supercond.*, vol. 15, no. 2 PART II, pp. 1763–6, Jun. 2005, doi: 10.1109/tasc.2005.849276.

[108] R. Soika, X. G. Garcia, and S. C. Nogales, "ENDESA Supercable, a 3.2 kA, 138 MVA, medium voltage superconducting power cable," *IEEE Trans. Appl. Supercond.*, vol. 21, no. 3 PART 2, pp. 972–5, Jun. 2011, doi: 10.1109/tasc.2010.2100357.

[109] S. S. Fetisov, V. V. Zubko, S. Yu Zanegin, A. A. Nosov, S. M. Ryabov, and V. S. Vysotsky, "Study of the First Russian Triaxial HTS Cable Prototypes," *IEEE Trans. Appl. Supercond.*, vol. 27, no. 4, pp. 1–5, Jun. 2017, doi: 10.1109/tasc.2017.2652854.

[110] G. Angeli, M. Bocchi, M. Ascade, V. Rossi, A. Valzasina, and L. Martini, "Development of Superconducting Devices for Power Grids in Italy: Update about the SFCL Project and Launching of the Research Activity on HTS Cables," *IEEE Trans. Appl. Supercond.*, vol. 27, no. 4, pp. 1–6, Jun. 2017, doi: 10.1109/tasc.2016.2639022.

[111] Y. J. Won *et al.*, “Installation Design of 23kV 50MVA class HTS Cable in South Korea,” *J. Phys.: Conf. Series*, vol. 1054, no. 1, p. 012074, Jul. 2018, doi: 10.1088/1742-6596/1054/1/012074.

[112] M. A. Neves *et al.*, “Development of the first brazilian project on superconducting power cable,” *IEEE Trans. Appl. Supercond.*, vol. 26, no. 3, pp. 1–5, Apr. 2016, doi: 10.1109/tasc.2016.2521582.

[113] W. Prusseit and R. Bach, “The Munich SuperLink project,” *Transformers Magazine*, 2021. [Online]. Available: https://transformers-magazine.com/files/The-Munich-Super-Link-project.pdf

[114] S. M, M. F, N. M, and H. A, “AmpaCity Project—Worldwide First Superconducting Cable and Fault Current Limiter Installation in a German City Center,” 2013.

[115] S. M, M. F, and N. M, “AmpaCity Project—Update on Worlds First Superconducting Cable and Fault Current Limiter Installation in a German City Center,” 2015.

[116] T. Ishigohka, “Possibility of DC underwater superconducting power cable cooled by long-length heat-pipe,” *Proceedings of the IEEE Power Engineering Society Transmission and Distribution Conference*, vol. 2, no. ASIA PACIFIC, pp. 1294–1297, 2002, doi: 10.1109/TDC.2002.1177666.

[117] H. Ohsaki, Z. Lv, N. Matsushita, M. Sekino, T. Koseki, and M. Tomita, “Superconducting power cable application in DC electric railway systems,” *IEEE Transactions on Applied Superconductivity*, vol. 23, no. 3, 2013, doi: 10.1109/TASC.2013.2237874.

[118] P. Cheetham, B. Darbha, S. Telikapalli, C. Kim, M. Coleman, and S. Pamidi, “Superconducting DC power distribution networks for electric aircraft,” *AIAA/IEEE Electric Aircraft Technologies Symposium (EATS)*, 2020.

[119] G. Lyu, Y. Terao, and H. Ohsaki, “HTS Cable Design for Aircraft Application Considering Fault Conditions,” *IEEE Transactions on Applied Superconductivity*, vol. 34, no. 3, pp. 1–5, May 2024, doi: 10.1109/TASC.2024.3356426.

[120] T. Izumi *et al.*, “Development of Superconducting Cable and Coated Conductors for Electric Propulsion System of Airplane,” *IEEE Transactions on Applied Superconductivity*, vol. 33, no. 5, Aug. 2023, doi: 10.1109/TASC.2023.3258902.

[121] Fujikura Company, “Introduction of Fujikura RE-based High-Temperature Superconductor,” Jun. 2022.

[122] SuperPower Inc., “2G HTS Wire Specification | SuperPower.” Accessed: Nov. 15, 2024. [Online]. Available: https://www.superpower-inc.com/specification.aspx

[123] F. W. Grover, *Inductance Calculations*. Dover Phoenix Editions, 2004.

[124] E. Shabagin, C. Heidt, S. Strauß, and S. Grohmann, “Modelling of 3D temperature profiles and pressure drop in concentric three-phase HTS power cables,” *Cryogenics (Guildf)*, vol. 81, pp. 24–32, Jan. 2017, doi: 10.1016/J.CRYOGENICS.2016.11.004.

[125] W. T. B. De Sousa, E. Shabagin, D. Kottonau, and M. Noe, “An open-source 2D finite difference based transient electro-thermal simulation model for three-phase concentric superconducting power cables,” *Supercond Sci Technol*, vol. 34, no. 1, Dec. 2020, doi: 10.1088/1361-6668/ABC2B0.

[126] P.J. Linstrom and W.G. Mallard (Eds.), *NIST Chemistry WebBook, NIST Standard Reference Database Number 69*. Gaithersburg MD, 20899: National Institute of Standards and Technology (NIST). [Online]. Available: https://doi.org/10.18434/T4D303

[127] C. Schaschke, “A Dictionary of Chemical Engineering,” *A Dictionary of Chemical Engineering*, Sep. 2014, doi: 10.1093/ACREF/9780199651450.001.0001.

[128] K. Kauder, “Strömungs- und Widerstandsverhalten in gewellten Rohren,” Hannover, 1971.

[129] E. Shabagin, “ Untersuchungen zum Druck- und Temperaturverhalten supraleitender Kabel,” Karlsruhe Institute of Technology, 2015.

[130] M. N. Özisik, “Heat transfer : a basic approach / M. Necati Özisik,” 1985, Accessed: May 17, 2023. [Online]. Available: https://books.google.com/books/about/Heat_Transfer.html?id=1-JSAAAAMAAJ

[131] D. W. Hahn and M. N. Özişik, “Heat Conduction: Third Edition,” *Heat Conduction: Third Edition*, Aug. 2012, doi: 10.1002/9781118411285.

[132] Y. Terao, A. Seta, H. Ohsaki, H. Oyori, and N. Morioka, “Lightweight Design of Fully Superconducting Motors for Electrical Aircraft Propulsion Systems,” *IEEE Transactions on Applied Superconductivity*, vol. 29, no. 5, Aug. 2019, doi: 10.1109/TASC.2019.2902323.

[133] “Siemens develops world-record electric motor for aircraft.” Accessed: Sep. 30, 2024. [Online]. Available: https://press.siemens.com/global/en/pressrelease/siemens-develops-world-record-electric-motor-aircraft

[134] Y. Terao, W. Kong, H. Ohsaki, H. Oyori, and N. Morioka, “Electromagnetic Design of Superconducting Synchronous Motors for Electric Aircraft Propulsion,” *IEEE Transactions on Applied Superconductivity*, vol. 28, no. 4, Jun. 2018, doi: 10.1109/TASC.2018.2823503.

[135] B. Łukasik, “Turboelectric Distributed Propulsion System As a Future Replacement for Turbofan Engines,” *Proceedings of the ASME Turbo Expo*, vol. 1, Aug. 2017, doi: 10.1115/GT2017-63834.

[136] P. J. Masson and C. A. Luongo, "High power density superconducting motor for all-electric aircraft propulsion," *IEEE Transactions on Applied Superconductivity*, vol. 15, no. 2 PART II, pp. 2226–2229, Jun. 2005, doi: 10.1109/TASC.2005.849618.

[137] C. D. Manolopoulos, M. F. Iacchetti, A. C. Smith, K. Berger, M. Husband, and P. Miller, "Stator Design and Performance of Superconducting Motors for Aerospace Electric Propulsion Systems," *IEEE Transactions on Applied Superconductivity*, vol. 28, no. 4, Jun. 2018, doi: 10.1109/TASC.2018.2814742.

[138] D. Dezhin, N. Ivanov, K. Kovalev, I. Kobzeva, and V. Semenihin, "System Approach of Usability of HTS Electrical Machines in Future Electric Aircraft," *IEEE Transactions on Applied Superconductivity*, vol. 28, no. 4, Jun. 2018, doi: 10.1109/TASC.2017.2787180.

[139] M. Zhang, F. Eastham, and W. Yuan, "Design and Modeling of 2G HTS Armature Winding for Electric Aircraft Propulsion Applications," *IEEE Transactions on Applied Superconductivity*, vol. 26, no. 3, Apr. 2016, doi: 10.1109/TASC.2016.2539548.

[140] M. S. Hassan, T. Asano, M. Shoyama, and G. M. Dousoky, "Performance Investigation of Power Inverter Components Submersed in Subcooled Liquid Nitrogen for Electric Aircraft," *Electronics 2022, Vol. 11, Page 826*, vol. 11, no. 5, p. 826, Mar. 2022, doi: 10.3390/ELECTRONICS11050826.

[141] R. Chen *et al.*, "A Cryogenically-Cooled MW Inverter for Electric Aircraft Propulsion," *2020 AIAA/IEEE Electric Aircraft Technologies Symposium (EATS)*, pp. 1–10, 2020, doi: 10.2514/6.2020-3598.

[142] K. Rajashekara and B. Akin, "A review of cryogenic power electronics-status and applications," *Proceedings of the 2013 IEEE International Electric Machines and Drives Conference, IEMDC 2013*, pp. 899–904, 2013, doi: 10.1109/IEMDC.2013.6556204.

[143] H. Gui *et al.*, "Review of Power Electronics Components at Cryogenic Temperatures," *IEEE Trans Power Electron*, vol. 35, no. 5, pp. 5144–5156, May 2020, doi: 10.1109/TPEL.2019.2944781.

[144] P. Haldar *et al.*, "Improving performance of cryogenic power electronics," *IEEE Transactions on Applied Superconductivity*, vol. 15, no. 2 PART II, pp. 2370–2375, Jun. 2005, doi: 10.1109/TASC.2005.849668.

[145] M. S. Hassan, T. Asano, M. Shoyama, and G. M. Dousoky, "Performance Investigation of Power Inverter Components Submersed in Subcooled Liquid Nitrogen for Electric Aircraft," *Electronics 2022, Vol. 11, Page 826*, vol. 11, no. 5, p. 826, Mar. 2022, doi: 10.3390/ELECTRONICS11050826.

[146] E. Dogmus *et al.*, "InAlGaN/GaN HEMTs at cryogenic temperatures," *Electronics (Switzerland)*, vol. 5, no. 2, Jun. 2016, doi: 10.3390/ELECTRONICS5020031.

[147] F. Wang *et al.*, "Mw-class cryogenically-cooled inverter for electric-aircraft applications," *AIAA Propulsion and Energy Forum and Exposition, 2019*, 2019, doi: 10.2514/6.2019-4473.

[148] K. Rajashekara and B. Akin, "Cryogenic Power Conversion Systems: The next step in the evolution of power electronics technology," *IEEE Electrification Magazine*, vol. 1, no. 2, pp. 64–73, 2013, doi: 10.1109/MELE.2013.2282195.

[149] E. K. Beser, "Electrical equivalent circuit for modelling permanent magnet synchronous motors," *Journal of Electrical Engineering*, vol. 72, no. 3, pp. 176–183, Jun. 2021, doi: 10.2478/JEE-2021-0024.

[150] A. Khonya, "AC-DC Converters for Medium Voltage Direct Current Networks with Integrated Renewable Energy Sources," Politecnico di Milano, Milan, 2021. Accessed: Sep. 21, 2024. [Online]. Available: https://www.politesi.polimi.it/handle/10589/174884.

[151] H. Ghanayem, M. Alathamneh, and R. M. Nelms, "PMSM Field-Oriented Control with Independent Speed and Flux Controllers for Continuous Operation under Open-Circuit Fault at Light Load Conditions," *Energies 2024, Vol. 17, Page 593*, vol. 17, no. 3, p. 593, Jan. 2024, doi: 10.3390/EN17030593.

[152] F. Blaschke, "The Principle of Field Orientation as Applied to the New Transvector Closed-Loop Control System for Rotating-Field Machine," *Siemens Review*, vol. 34, no. 6, pp. 217–220, 1972, doi: 10.1541/IEEJIAS.116.652.

[153] F. Yusivar, N. Hidayat, R. Gunawan, and A. Halim, "Implementation of field oriented control for permanent magnet synchronous motor," *Proceedings of 2014 International Conference on Electrical Engineering and Computer Science, ICEECS 2014*, pp. 359–362, Feb. 2014, doi: 10.1109/ICEECS.2014.7045278.

[154] Q. Song and C. Jia, "Robust Speed Controller Design for Permanent Magnet Synchronous Motor Drives Based on Sliding Mode Control," *Energy Procedia*, vol. 88, pp. 867–873, Jun. 2016, doi: 10.1016/J.EGYPRO.2016.06.102.

[155] K. C. Odo, S. V. Egoigwe, and C. U. Ogbuka, "(PDF) A Model-based PI Controller Tuning and Design for Field Oriented Current Control of Permanent Magnet Synchronous Motor," *IOSR Journal of Electrical and Electronics Engineering (IOSR-JEEE)*, Aug. 2019, Accessed: Oct. 02, 2024. [Online]. Available: https://www.researchgate.net/publication/335444205_A_Model-based_PI_Controller_Tuning_and_Design_for_Field_Oriented_Current_Control_of_Permanent_Magnet_Synchronous_Motor

[156] M. Abassi, A. Khlaief, O. Saadaoui, A. Chaari, and M. Boussak, "Performance analysis of FOC and DTC for PMSM drives using SVPWM technique," *16th International*

Conference on Sciences and Techniques of Automatic Control and Computer Engineering, STA 2015, pp. 228–233, Jul. 2016, doi: 10.1109/STA.2015.7505167.

[157] F. Korkmaz, I. Topaloğlu, M. F. Çakir, and R. Gürbüz, "Comparative performance evaluation of FOC and DTC controlled PMSM drives," *International Conference on Power Engineering, Energy and Electrical Drives*, pp. 705–708, 2013, doi: 10.1109/POWERENG.2013.6635696.

[158] S. K. Dwivedi, M. Laursen, and S. Hansen, "Voltage vector based control for PMSM in industry applications," *IEEE International Symposium on Industrial Electronics*, pp. 3845–3850, 2010, doi: 10.1109/ISIE.2010.5637742.

[159] I. Qureshi and V. Sharma, "Analysis of different control schemes of PMSM motor and also a comparison of FOPI and PI controller for sensorless MSVPWMM scheme," *e-Prime - Advances in Electrical Engineering, Electronics and Energy*, vol. 6, p. 100359, Dec. 2023, doi: 10.1016/J.PRIME.2023.100359.

[160] A. Flah, I. A. Khan, A. Agarwal, L. Sbita, and M. G. Simoes, "Field-oriented control strategy for double-stator single-rotor and double-rotor single-stator permanent magnet machine: Design and operation," *Computers & Electrical Engineering*, vol. 90, p. 106953, Mar. 2021, doi: 10.1016/J.COMPELECENG.2020.106953.

[161] K. Mathew K, D. M. Abraham, and A. Harish, "Speed regulation of PMSM drive in electric vehicle applications with sliding mode controller based on harris Hawks optimization," *e-Prime - Advances in Electrical Engineering, Electronics and Energy*, vol. 9, p. 100643, Sep. 2024, doi: 10.1016/J.PRIME.2024.100643.

[162] P. Krause, O. Wasynczuk, S. Sudhoff, and S. Pekarek, *Analysis of Electric Machinery and Drive Systems*. Piscatawy, NJ: Wiley-IEEE Press, 2013. Accessed: Sep. 20, 2024. [Online]. Available: https://ieeexplore.ieee.org/book/6712180

[163] "FF300R06KE3 | 600 V, 300 A dual IGBT module - Infineon Technologies." [Online]. Available: https://www.infineon.com/cms/en/product/power/igbt/igbt-modules/ff300r06ke3/

9 Appendices

A.1 HTS Tape Materials Data

The intrinsic properties of the materials used in the superconducting tapes are given in Table 9.1 based on the equation (3.6) in Chapter 3. In this table, T is temperature, and the formula shown for REBCO resistivity is when it is normal state.

Table 9.1: Intrinsic properties of the materials used in the HTS tapes and DC/AC cables [92].

Material	Specific Resistivity (ρ) for 75 K<T<300 K [Ω.m]	Specific Heat Capacity (c_p) for 75 K<T<300 K [J/(kg.K)]	Density (ρ) [kg/m³]
Silver (Ag)	$6.17 \times 10^{-11}T - 2.082 \times 10^{-9}$	$220.5 + 0.046T$	1049
REBCO	$1 \times 10^{-4}T - 10 \times 10^{-4}$ ($T > T_c$)	$390\left(\frac{250}{T}\right)^2 \frac{e^{\frac{250}{T}}}{\left(e^{\frac{250}{T}} - 1\right)^2}$	6390
Hastelloy (Hy)	$8.958 \times 10^{-11}T + 1.103 \times 10^{-6}$	$190 + 388\left(1 - e^{-\frac{T}{310}}\right)$	8890
Copper (Cu)	$6.841 \times 10^{-11}T - 3.06 \times 10^{-9}$	$390.9 - 593.4e^{-0.014T}$	8960
Aluminum (Al)	$1.13 \times 10^{-10}T - 6.718 \times 10^{-9}$	$6 \times 10^{-5}T^3 - 0.0471T^2 + 13.281T - 423.27$	2700

A.2 HTS Tapes Heat Capacity Calculation

The heat capacity of a layer in the HTS tape, C_{layer}, is calculated from its specific heat capacity $c_{\mathrm{p,layer}}$ and mass m_{layer} with equation (9.1) based on the data given in Table 9.1.

$$C_{\mathrm{layer}} = m_{\mathrm{layer}} \cdot c_{\mathrm{p,layer}} \tag{9.1}$$

The layer mass m_{layer} is calculated using equation (9.2) with the layer density ρ_{layer}, cross-section area A_{layer}, and length which is equal to tape length l_{Tape}.

$$m_{\mathrm{layer}} = \rho_{\mathrm{layer}} \cdot A_{\mathrm{layer}} \cdot l_{\mathrm{Tape}} \tag{9.2}$$

Finally, by determining the heat capacities of all N_{layer} layers, the total heat capacity of the tape can be computed using equation (9.3).

$$C_{\mathrm{Tape}} = \sum_{i=1}^{N_{\mathrm{layer}}} C_{\mathrm{layer,i}} \tag{9.3}$$

A.3 Liquid Nitrogen Properties

The parameters α_i to calculate $h_{\mathrm{c,LN_2}}$ which was provided in Chapter 3, equation (3.13) are given below:

$$\alpha_0 = 38147, \alpha_1 = -28209, \alpha_2 = 7232.8,$$
$$\alpha_3 = -494.38, \alpha_4 = 13.25, \alpha_5 = -0.12773 \tag{9.4}$$

The LN2 properties at a pressure of 15 bar and a temperature of 77 K are as below [126]:

Table 9.2: LN2 properties at a pressure of 15 bar and a temperature of 77 K [126].

Parameter	Symbol	Value
Desnity	$\rho_{\mathrm{LN_2}}$	811 kg/m^3
Viscosity	$\eta_{\mathrm{LN_2}}$	166.68 μPa.s
Specific Heat Capacity	$c_{\mathrm{p,LN_2}}$	2025 J/(kg.K)
Thermal Conductivity	$k_{\mathrm{LN_2}}$	0.15 W/(m.K)

A.4 HTS Cable Layers Properties

In Table 9.3, the intrinsic properties of the cable layers are detailed.

Table 9.3: Intrinsic properties of the materials used in the HTS cable [92], [125].

Cable Layer	Thermal Conductivity (k) for 75 K<T<300 K [W/(m.K)]	Specific Heat Capacity (c_p) for 75 K<T<300 K [J/(kg.K)]	Density (ρ) [kg/m^3]
Core/Shield Copper	534.2	See Table 9.1.	8960
HTS Tape	150	Calculated based on the different layers data given in Table 9.1 and section A.1.	Calculated based on the different layers data given in Table 9.1.**
Polypropylene Laminated Paper (PPLP)	0.15*	430	8890
Liquid Nitrogen (LN_2)	0.15	2025	811

* Immersed in LN_2.

** Density of the HTS tapes is calculated based on the density, mass, and volume of the sub-layers in the tapes. Their properties are described in Table 9.1.

A.5 Cable Cooling Fluid Calculations

The following procedure shows the calculation of the cooling fluid Reynolds number Re_{Fld}.

First, from fluid mass flow rate $\dot{m}_{\mathrm{Fld}}$, density ρ_{Fld}, and the fluid duct area A_{Fld}, the fluid velocity in the duct v_{Fld} is calculated with equation (9.5):

$$v_{\mathrm{Fld}} = \frac{\dot{m}_{\mathrm{Fld}}}{\rho_{\mathrm{Fld}} \cdot A_{\mathrm{Fld}}} \tag{9.5}$$

Then, a base Reynolds number $Re_{\mathrm{Fld,b}}$ needs to be calculated with equation (9.6) [129]. In this equation, $d_{\mathrm{H,Fld}}$ is the fluid duct hydraulic diameter, and μ_{Fld} is the fluid dynamic viscosity.

$$Re_{\mathrm{Fld,b}} = \frac{v_{\mathrm{Fld}} \cdot d_{\mathrm{H,Fld}} \cdot \rho_{\mathrm{Fld}}}{\mu_{\mathrm{Fld}}} \tag{9.6}$$

This equation is valid for the case of an inner fluid channel, but for an outside channel, it needs to be adjust based on the ratio between the inner and outer fluid duct diameters, represented by d_{in} and d_{out}, which is named α and defined as follows:

$$\alpha = \frac{d_{\mathrm{in}}}{d_{\mathrm{out}}} \tag{9.7}$$

It is worth to mention that in the particular cable considered in this work, the shield copper outer diameter $d_{\mathrm{Sh,Out}}$ is equal to the fluid duct inner diameter, so these parameters can be replaced with each other.

With the value of α, the accurate value of Reynolds number Re_{Fld} for the outer channel (which is the case of the considered cable in this work) can be calculated with the following equation:

$$Re_{\mathrm{Fld}} = Re_{\mathrm{Fld,b}} \cdot \frac{(1+\alpha^2) \cdot \ln(\alpha) + (1-\alpha^2)}{(1-\alpha)^2 \cdot \ln(\alpha)} \tag{9.8}$$

To calculate a coolant convective heat coefficient in a cable, the approach below must be followed.

First, the fluid Prandtl number Pr_{Fld} needs to be calculated with equation (9.9) from fluid specific heat capacity $c_{\mathrm{p,Fld}}$, viscosity μ_{Fld}, and thermal conductivity k_{Fld}.

$$Pr_{\mathrm{Fld}} = \frac{\mu_{\mathrm{Fld}} \cdot c_{\mathrm{p,Fld}}}{k_{\mathrm{Fld}}} \tag{9.9}$$

Then, based on the fluid Reynolds number Re_{Fld} which was calculated in equation (9.8), the coolant duct friction coefficient ξ_{Fld} is calculated as below:

$$\xi_{\mathrm{Fld}} = (1.8 \log_{10}(Re_{\mathrm{Fld}}) - 1.5)^{-2} \tag{9.10}$$

In the next step, the fluid Nusselt number Nu_{Fld} needs to be calculated with the following equation [129]:

$$Nu_{\mathrm{Fld}} = \frac{\left(\frac{\xi_{\mathrm{Fld}}}{8}\right) \cdot Re_{\mathrm{Fld}} \cdot Pr_{\mathrm{Fld}}}{k_1 + 12.7\left(\frac{\xi_{\mathrm{Fld}}}{8}\right)^{\frac{1}{2}} \cdot \left(Pr_{\mathrm{Fld}}^{\frac{2}{3}} - 1\right)} \cdot k_{\mathrm{Nu,c}} \tag{9.11}$$

In this equation, $k_{\mathrm{Nu,c}}$ is the Nusselt number correction factor which depends on the ratio of the inner and outer duct diameters α (equation (9.7)), and k_1 is a self-built parameter. The parameters $k_{\mathrm{Nu,c}}$ and k_1 are defined with equations (9.12) and (9.13), respectively.

$$k_{\mathrm{Nu,c}} = \frac{0.75\alpha^{-0.17} + (0.9 - 0.15\alpha^{0.6})}{1 + \alpha} \tag{9.12}$$

$$k_1 = 1.07 + \frac{900}{Re_{\mathrm{Fld}}} - \frac{0.63}{(1 + 10Pr_{\mathrm{Fld}})} \tag{9.13}$$

Finally, the fluid convective heat coefficient $h_{\mathrm{c,Fld}}$ is calculated with equation (9.14):

$$h_{\mathrm{c,Fld}} = \frac{Nu_{\mathrm{Fld}} \cdot k_{\mathrm{Fld}}}{d_{\mathrm{H,Fld}}} \tag{9.14}$$

A.6 Additional Remarks on the 1-D, Transient, FDM Based Model of the HTS DC Cable

Apart from the FDM based heat transfer equation for an element m in solid layers which was described in Chapter 4, section 4.6.2, equation (4.24), the equations for the elements m=1 and m=M are given below.

For the element m=1:

$$[\psi_1 + 4] \cdot T_1^{z+1} - 4T_2^{z+1} = \psi_1 \cdot T_1^z + \frac{(\Delta r)^2}{k_1} \cdot g_1^{z+1} \tag{9.15}$$

For the element m=M:

$$\begin{aligned} -2T_{M-1}^{z+1} + \left[\psi_M + 2 + \varphi \cdot \left(\frac{1}{2M} + 1\right)\right] \cdot T_M^{z+1} + \left[-\varphi \cdot \left(\frac{1}{2M} + 1\right)\right] \cdot T_{Fld}^{z+1} \\ = \psi_M \cdot T_M^z + \frac{(\Delta r)^2}{k_M} \cdot g_M^{z+1} \end{aligned} \tag{9.16}$$

The matrices of the equation (4.28) in Chapter 4, section 4.5.2 are shown in equations (9.17) and (9.18).

$$T = \begin{bmatrix} T_1 \\ T_2 \\ \vdots \\ T_{m-1} \\ T_m \\ T_{m+1} \\ \vdots \\ T_{M-1} \\ T_M \\ T_{Fld} \end{bmatrix}, \quad \psi = \begin{bmatrix} \psi_1 \\ \psi_2 \\ \vdots \\ \psi_{m-1} \\ \psi_m \\ \psi_{m+1} \\ \vdots \\ \psi_{M-1} \\ \psi_M \\ \psi_{Fld} \end{bmatrix}, \quad B = \begin{bmatrix} \frac{(\Delta r)^2}{k_1} \cdot g_1^{z+1} \\ \frac{(\Delta r)^2}{k_2} \cdot g_2^{z+1} \\ \vdots \\ \frac{(\Delta r)^2}{k_{m-1}} \cdot g_{m-1}^{z+1} \\ \frac{(\Delta r)^2}{k_m} \cdot g_m^{z+1} \\ \frac{(\Delta r)^2}{k_{m+1}} \cdot g_{m+1}^{z+1} \\ \vdots \\ \frac{(\Delta r)^2}{k_{M-1}} \cdot g_{M-1}^{z+1} \\ \frac{(\Delta r)^2}{k_M} \cdot g_M^{z+1} \\ q_{rad} \end{bmatrix} \tag{9.17}$$

$$A = \begin{bmatrix} \psi_1 + 4 & -4 & 0 & \cdots & 0 & 0 & 0 & \cdots & 0 & 0 \\ A_{2,1} & A_{2,2} & A_{2,3} & \cdots & 0 & 0 & 0 & \cdots & 0 & 0 \\ \vdots & \vdots & \vdots & \ddots & \vdots & \vdots & \vdots & \ddots & \vdots & \vdots \\ 0 & 0 & 0 & \cdots & A_{m-1,m-1} & A_{m-1,m} & 0 & \cdots & 0 & 0 \\ 0 & 0 & 0 & \cdots & A_{m,m-1} & A_{m,m} & A_{m,m+1} & \cdots & 0 & 0 \\ 0 & 0 & 0 & \cdots & 0 & A_{m+1,m} & A_{m+1,m+1} & \cdots & 0 & 0 \\ \vdots & \vdots & \vdots & \ddots & \vdots & \vdots & \vdots & \ddots & \vdots & \vdots \\ 0 & 0 & 0 & \cdots & 0 & 0 & 0 & \cdots & A_{M-1,M} & 0 \\ 0 & 0 & 0 & \cdots & 0 & 0 & 0 & \cdots & A_{M,M} & A_{M,N_r} \\ 0 & 0 & 0 & \cdots & 0 & 0 & 0 & \cdots & A_{N_r,M} & A_{N_r,N_r} \end{bmatrix}$$

$$A_{m,m-1} = \left(\frac{1}{2m} - 1\right) \cdot \left(\frac{2}{1 + \eta_{m,m-1}}\right) \tag{9.18}$$

$$A_{m,m} = \psi_m + 2 + \left(\frac{1}{2m} - 1\right) \cdot \left(\frac{\eta_{m,m-1} - 1}{\eta_{m,m-1} + 1}\right) + \left(\frac{1}{2m} + 1\right) \cdot \left(\frac{\eta_{m+1,m} - 1}{\eta_{m+1,m} + 1}\right)$$

$$A_{m,m+1} = \left(\frac{1}{2m} + 1\right) \cdot \left(\frac{-2\eta_{m+1,m}}{1 + \eta_{m+1,m}}\right)$$

$$A_{M,M-1} = -2\,, A_{M,M} = \psi_M + 2 + \varphi \cdot \left(\frac{1}{2M} + 1\right)\,, A_{M,N_r} = -\varphi \cdot \left(\frac{1}{2M} + 1\right)$$

$$A_{N_r,M} = -h_{c,Fld} \cdot Pe_{Fld,in} \quad , \quad A_{N_r,N_r} = \psi_{Fld} + h_{c,Fld} \cdot Pe_{Fld,in}$$

A.7 Additional Remarks on the 2-D, Stationary, FDM Based Model of the HTS DC Cable

Apart from the FDM based heat transfer equation for an element m in solid layers which was described in Chapter 4, section 4.6.2, equation (4.24), the equations for the elements m=1 and m=M are given below.

For the element m=1:

$$4T_{1,\mathrm{n}} - 4T_{2,\mathrm{n}} = \frac{(\Delta r)^2}{k_1} \cdot g_{1,\mathrm{n}} \tag{9.19}$$

For the element m=M:

$$\begin{aligned} -2T_{\mathrm{M-1,n}} + \left[2 + \varphi \cdot \left(\frac{1}{2M} + 1\right)\right] T_{\mathrm{M,n}} + \left[-\varphi \cdot \left(\frac{1}{2M} + 1\right)\right] \cdot T_{\mathrm{Fld,n}} \\ = \frac{(\Delta r)^2}{k_{\mathrm{M}}} \cdot g_{\mathrm{M,n}} \end{aligned} \tag{9.20}$$

The matrices of the equation (4.35) in Chapter 4, section 4.6.1 are shown in equations (9.21) and (9.22).

$$T_{\mathrm{n}} = \begin{bmatrix} T_{1,\mathrm{n}} \\ T_{2,\mathrm{n}} \\ \vdots \\ T_{\mathrm{m-1,n}} \\ T_{\mathrm{m,n}} \\ T_{\mathrm{m+1,n}} \\ \vdots \\ T_{\mathrm{M-1,n}} \\ T_{\mathrm{M,n}} \\ T_{\mathrm{Fld,n}} \end{bmatrix}, \quad \mathrm{B_n} = \begin{bmatrix} \frac{(\Delta r)^2}{k_1} \cdot g_{1,\mathrm{n}} \\ \frac{(\Delta r)^2}{k_2} \cdot g_{2,\mathrm{n}} \\ \vdots \\ \frac{(\Delta r)^2}{k_{\mathrm{m-1}}} \cdot g_{\mathrm{m-1,n}} \\ \frac{(\Delta r)^2}{k_{\mathrm{m}}} \cdot g_{\mathrm{m,n}} \\ \frac{(\Delta r)^2}{k_{\mathrm{m+1}}} \cdot g_{\mathrm{m+1,n}} \\ \vdots \\ \frac{(\Delta r)^2}{k_{\mathrm{M-1}}} \cdot g_{\mathrm{M-1,n}} \\ \frac{(\Delta r)^2}{k_{\mathrm{M}}} \cdot g_{\mathrm{M,n}} \\ \left[\frac{\dot{m}_{\mathrm{Fld}} \cdot c_{\mathrm{p,Fld}}}{\Delta x}\right] \cdot T_{\mathrm{Fld,n-1}} + q_{\mathrm{rad}} \end{bmatrix} \tag{9.21}$$

$$\mathrm{A_n} = \begin{bmatrix} 4 & -4 & 0 & \cdots & 0 & 0 & 0 & \cdots & 0 & 0 \\ -\frac{3}{4} & 2 & -\frac{5}{4} & \cdots & 0 & 0 & 0 & \cdots & 0 & 0 \\ \vdots & \vdots & \vdots & \ddots & \vdots & \vdots & \vdots & \ddots & \vdots & \vdots \\ 0 & 0 & 0 & \cdots & A_{\mathrm{m-1,m-1}} & A_{\mathrm{m-1,m}} & 0 & \cdots & 0 & 0 \\ 0 & 0 & 0 & \cdots & A_{\mathrm{m,m-1}} & A_{\mathrm{m,m}} & A_{\mathrm{m,m+1}} & \cdots & 0 & 0 \\ 0 & 0 & 0 & \cdots & 0 & A_{\mathrm{m+1,m}} & A_{\mathrm{m+1,m+1}} & \cdots & 0 & 0 \\ \vdots & \vdots & \vdots & \ddots & \vdots & \vdots & \vdots & \ddots & \vdots & \vdots \\ 0 & 0 & 0 & \cdots & 0 & 0 & 0 & \cdots & A_{\mathrm{M-1,M}} & 0 \\ 0 & 0 & 0 & \cdots & 0 & 0 & 0 & \cdots & A_{\mathrm{M,M}} & A_{\mathrm{M,N_r}} \\ 0 & 0 & 0 & \cdots & 0 & 0 & 0 & \cdots & A_{\mathrm{N_r,M}} & A_{\mathrm{N_r,N_r}} \end{bmatrix} \tag{9.22}$$

$$A_{\mathrm{N_r,M}} = -h_{\mathrm{c,Fld}} \cdot Pe_{\mathrm{Fld,in}} \quad , \quad A_{\mathrm{N_r,N_r}} = h_{\mathrm{c,Fld}} \cdot Pe_{\mathrm{Fld,in}} + \frac{\dot{m}_{\mathrm{Fld}} \cdot c_{\mathrm{p,Fld}}}{\Delta x}$$

A.8 Additional Remarks on the 2-D, Transient, FDM Based Model of the HTS DC Cable

The additional equations in the two half time-steps of the ADI method applied to the 2-D, transient, FDM modeling of the cable are given below.

First Half Time-step:

For the element m=1 in the axial element n:

$$\begin{aligned}-(\Delta r)^2 \cdot T_{1,n-1}^{z+\frac{1}{2}} + [\psi_1 + 2(\Delta r)^2] \cdot T_{1,n}^{z+\frac{1}{2}} - (\Delta r)^2 \cdot T_{1,n+1}^{z+\frac{1}{2}} \\ = [\psi_1 - 4(\Delta x)^2] \cdot T_{1,n}^{z} + 4(\Delta x)^2 \cdot T_{2,n}^{z} + \frac{(\Delta x \cdot \Delta r)^2}{k_1} \cdot g_{1,n}^{z+1}\end{aligned} \tag{9.23}$$

For the element m=M in the axial element n:

$$\begin{aligned}-(\Delta r)^2 \cdot T_{M,n-1}^{z+\frac{1}{2}} + [\psi_M + 2(\Delta r)^2] \cdot T_{M,n}^{z+\frac{1}{2}} - (\Delta r)^2 \cdot T_{M,n+1}^{z+\frac{1}{2}} \\ = 2(\Delta x)^2 \cdot T_{M-1,n}^{z} + \left[\psi_M - 2(\Delta x)^2 - \varphi \cdot (\Delta x)^2 \cdot \left(\frac{1}{2M} + 1\right)\right] \\ \cdot T_{M,n}^{z} + \left[\varphi \cdot (\Delta x)^2 \cdot \left(\frac{1}{2M} + 1\right)\right] \cdot T_{Fld,n}^{z} + \frac{(\Delta x \cdot \Delta r)^2}{k_M} \cdot g_{M,n}^{z+1}\end{aligned} \tag{9.24}$$

For the cooling fluid element m=N_r in the axial element n:

$$\begin{aligned}\left[-\frac{\dot{m}_{Fld} \cdot c_{p,Fld}}{\Delta x}\right] \cdot T_{Fld,n-1}^{z+\frac{1}{2}} + \left[\psi_{Fld} + \frac{\dot{m}_{Fld} \cdot c_{p,Fld}}{\Delta x}\right] \cdot T_{Fld,n}^{z+\frac{1}{2}} \\ = \left[h_{c,Fld,n} \cdot Pe_{Fld,in}\right] \cdot T_{M,n}^{z} + \left[\psi_{Fld} - h_{c,Fld,n} \cdot Pe_{Fld,in}\right] \\ \cdot T_{Fld,n}^{z} + q_{rad}\end{aligned} \tag{9.25}$$

In equation (9.25), the parameter ψ_{Fld} is self-defined as follows:

$$\psi_{Fld} = \frac{2\rho_{Fld} \cdot c_{p,Fld} \cdot A_{Fld}}{\Delta t} \tag{9.26}$$

The matrices for the axial element n in this model are detailed as below.

$$Tx_m^{z+\frac{1}{2}} = \begin{bmatrix} T_{m,1}^{z+\frac{1}{2}} \\ \vdots \\ T_{m,n}^{z+\frac{1}{2}} \\ \vdots \\ T_{m,N_x}^{z+\frac{1}{2}} \end{bmatrix} \quad , \quad Bx_m = \begin{bmatrix} \frac{(\Delta x \cdot \Delta r)^2}{k_m} \cdot g_{m,1}^{z+1} \\ \vdots \\ \frac{(\Delta x \cdot \Delta r)^2}{k_m} \cdot g_{m,n}^{z+1} \\ \vdots \\ \frac{(\Delta x \cdot \Delta r)^2}{k_m} \cdot g_{m,N_x}^{z+1} \end{bmatrix} \tag{9.27}$$

$$
Qr_{\mathrm{m}}^{\mathrm{z}} = \begin{bmatrix} F_{\mathrm{r},1} \cdot T_{\mathrm{m-1,1}}^{\mathrm{z}} + F_{\mathrm{r},2} \cdot T_{\mathrm{m,1}}^{\mathrm{z}} + F_{\mathrm{r},3} \cdot T_{\mathrm{m+1,1}}^{\mathrm{z}} \\ \vdots \\ F_{\mathrm{r},1} \cdot T_{\mathrm{m-1,n}}^{\mathrm{z}} + F_{\mathrm{r},2} \cdot T_{\mathrm{m,n}}^{\mathrm{z}} + F_{\mathrm{r},3} \cdot T_{\mathrm{m+1,n}}^{\mathrm{z}} \\ \vdots \\ F_{\mathrm{r},1} \cdot T_{\mathrm{m-1,N_x}}^{\mathrm{z}} + F_{\mathrm{r},2} \cdot T_{\mathrm{m,N_x}}^{\mathrm{z}} + F_{\mathrm{r},3} \cdot T_{\mathrm{m+1,N_x}}^{\mathrm{z}} \end{bmatrix}
$$

$$
F_{\mathrm{r},1} = (\Delta x)^2 \cdot \left(\frac{1}{2m} - 1\right) \cdot \left(\frac{-2}{1 + \eta_{\mathrm{m,m-1}}}\right)
$$

$$
F_{\mathrm{r},2} = \psi_{\mathrm{m,n}} - 2(\Delta x)^2 + (\Delta x)^2 \cdot \left(\frac{1}{2m} - 1\right) \cdot \left(\frac{1 - \eta_{\mathrm{m,m-1}}}{1 + \eta_{\mathrm{m,m-1}}}\right) + (\Delta x)^2 \cdot \left(\frac{1}{2m} + 1\right) \cdot \left(\frac{1 - \eta_{\mathrm{m+1,m}}}{1 + \eta_{\mathrm{m+1,m}}}\right)
$$

$$
F_{\mathrm{r},3} = (\Delta x)^2 \cdot \left(\frac{1}{2m} + 1\right) \cdot \left(\frac{2\eta_{\mathrm{m+1,m}}}{1 + \eta_{\mathrm{m+1,m}}}\right)
$$

$$
Ax_{\mathrm{m}} = \begin{bmatrix}
F_{\mathrm{x},1} & 2F_{\mathrm{x},2} & \cdots & 0 & 0 & 0 & \cdots & 0 & 0 \\
F_{\mathrm{x},2} & F_{\mathrm{x},1} & \cdots & 0 & 0 & 0 & \cdots & 0 & 0 \\
\vdots & \vdots & \ddots & \vdots & \vdots & \vdots & \ddots & \vdots & \vdots \\
0 & 0 & \cdots & F_{\mathrm{x},1} & F_{\mathrm{x},2} & 0 & \cdots & 0 & 0 \\
0 & 0 & \cdots & F_{\mathrm{x},2} & F_{\mathrm{x},1} & F_{\mathrm{x},2} & \cdots & 0 & 0 \\
0 & 0 & \cdots & 0 & F_{\mathrm{x},2} & F_{\mathrm{x},1} & \cdots & 0 & 0 \\
\vdots & \vdots & \ddots & \vdots & \vdots & \vdots & \ddots & \vdots & \vdots \\
0 & 0 & \cdots & 0 & 0 & 0 & \cdots & F_{\mathrm{x},1} & F_{\mathrm{x},2} \\
0 & 0 & \cdots & 0 & 0 & 0 & \cdots & 2F_{\mathrm{x},2} & F_{\mathrm{x},1}
\end{bmatrix}
$$

$$
F_{\mathrm{x},1} = \psi_{\mathrm{m}} + 2(\Delta r)^2 \quad , \quad F_{\mathrm{x},2} = -(\Delta r)^2
$$

Second Half Time-step:

For the element m=1 in the axial element n:

$$
\begin{aligned}
&[\psi_1 + 4(\Delta x)^2] \cdot T_{1,\mathrm{n}}^{\mathrm{z}+1} - 4(\Delta x)^2 \cdot T_{2,\mathrm{n}}^{\mathrm{z}+1} \\
&\quad = (\Delta r)^2 \cdot T_{1,\mathrm{n-1}}^{\mathrm{z}+\frac{1}{2}} + [\psi_1 - 2(\Delta r)^2] \cdot T_{1,\mathrm{n}}^{\mathrm{z}+\frac{1}{2}} + (\Delta r)^2 \cdot T_{1,\mathrm{n+1}}^{\mathrm{z}+\frac{1}{2}} \\
&\quad + \frac{(\Delta x \cdot \Delta r)^2}{k_1} \cdot g_{1,\mathrm{n}}^{\mathrm{z}+1}
\end{aligned} \tag{9.28}
$$

For the element m=M in the axial element n:

$$
\begin{aligned}
&-2(\Delta x)^2 \cdot T_{\mathrm{M-1,n}}^{\mathrm{z}} + \left[\psi_{\mathrm{M}} + 2(\Delta x)^2 + \varphi \cdot (\Delta x)^2 \cdot \left(\frac{1}{2M} + 1\right)\right] \cdot T_{\mathrm{M,n}}^{\mathrm{z}} \\
&\quad + \left[-\varphi \cdot (\Delta x)^2 \cdot \left(\frac{1}{2M} + 1\right)\right] \cdot T_{\mathrm{Fld,n}}^{\mathrm{z}} \\
&\quad = (\Delta r)^2 \cdot T_{\mathrm{M,n-1}}^{\mathrm{z}+\frac{1}{2}} + [\psi_{\mathrm{M}} - 2(\Delta r)^2] \cdot T_{\mathrm{M,n}}^{\mathrm{z}+\frac{1}{2}} + (\Delta r)^2 \cdot T_{\mathrm{M,n+1}}^{\mathrm{z}+\frac{1}{2}} \\
&\quad + \frac{(\Delta x \cdot \Delta r)^2}{k_{\mathrm{M}}} \cdot g_{\mathrm{M,n}}^{\mathrm{z}+1}
\end{aligned}
\tag{9.29}
$$

For the cooling fluid element $m = N_{\mathrm{r}}$ in the axial element n:

$$
\begin{aligned}
&\left[-h_{\mathrm{c,Fld,n}} \cdot Pe_{\mathrm{Fld,in}}\right] \cdot T_{\mathrm{M,n}}^{\mathrm{z}+1} + \left[\psi_{\mathrm{Fld}} + h_{\mathrm{c,Fld,n}} \cdot Pe_{\mathrm{Fld,in}}\right] \cdot T_{\mathrm{Fld,n}}^{\mathrm{z}+1} \\
&\quad = \left[\frac{\dot{m}_{\mathrm{Fld}} \cdot c_{\mathrm{p,Fld}}}{\Delta x}\right] \cdot T_{\mathrm{Fld,n-1}}^{\mathrm{z}+\frac{1}{2}} + \left[\psi_{\mathrm{Fld}} - \frac{\dot{m}_{\mathrm{Fld}} \cdot c_{\mathrm{p,Fld}}}{\Delta x}\right] \cdot T_{\mathrm{Fld,n}}^{\mathrm{z}+\frac{1}{2}} \\
&\quad + q_{\mathrm{rad}}
\end{aligned}
\tag{9.30}
$$

The matrices for the axial element n in this model are detailed as below.

$$
r_{\mathrm{n}}^{\mathrm{z}+1} = \begin{bmatrix} T_{\mathrm{1,n}}^{\mathrm{z}+1} \\ \vdots \\ T_{\mathrm{m,n}}^{\mathrm{z}+1} \\ \vdots \\ T_{\mathrm{Fld,n}}^{\mathrm{z}+1} \end{bmatrix}, \qquad Br_{\mathrm{n}} = \begin{bmatrix} \frac{(\Delta x \cdot \Delta r)^2}{k_{\mathrm{m}}} \cdot g_{\mathrm{1,n}}^{\mathrm{z}+1} \\ \vdots \\ \frac{(\Delta x \cdot \Delta r)^2}{k_{\mathrm{m}}} \cdot g_{\mathrm{m,n}}^{\mathrm{z}+1} \\ \vdots \\ \frac{(\Delta x \cdot \Delta r)^2}{k_{\mathrm{m}}} \cdot g_{\mathrm{Fld,n}}^{\mathrm{z}+1} \end{bmatrix}
$$

$$
Qx_{\mathrm{n}}^{\mathrm{z}+\frac{1}{2}} = \begin{bmatrix} H_{\mathrm{x,1}} \cdot T_{\mathrm{1,n-1}}^{\mathrm{z}+\frac{1}{2}} + H_{\mathrm{x,2}} \cdot T_{\mathrm{1,n}}^{\mathrm{z}+\frac{1}{2}} + H_{\mathrm{x,3}} \cdot T_{\mathrm{1,n+1}}^{\mathrm{z}+\frac{1}{2}} \\ \vdots \\ H_{\mathrm{x,1}} \cdot T_{\mathrm{m,n-1}}^{\mathrm{z}+\frac{1}{2}} + H_{\mathrm{x,2}} \cdot T_{\mathrm{m,n}}^{\mathrm{z}+\frac{1}{2}} + H_{\mathrm{x,3}} \cdot T_{\mathrm{m,n+1}}^{\mathrm{z}+\frac{1}{2}} \\ \vdots \\ H_{\mathrm{x,4}} \cdot T_{\mathrm{Fld,n-1}}^{\mathrm{z}+\frac{1}{2}} + H_{\mathrm{x,5}} \cdot T_{\mathrm{Fld,n}}^{\mathrm{z}+\frac{1}{2}} \end{bmatrix} \tag{9.31}
$$

$$
H_{\mathrm{x,1}} = (\Delta r)^2, H_{\mathrm{x,2}} = \psi_{\mathrm{m,n}} - 2(\Delta r)^2, H_{\mathrm{x,3}} = (\Delta r)^2
$$

$$
H_{\mathrm{x,4}} = \frac{\dot{m}_{\mathrm{Fld}} \cdot c_{\mathrm{p,Fld}}}{\Delta x}, H_{\mathrm{x,5}} = \psi_{\mathrm{Fld}} - \frac{\dot{m}_{\mathrm{Fld}} \cdot c_{\mathrm{p,Fld}}}{\Delta x}
$$

$$
Ar_{\mathrm{n}} = \begin{bmatrix}
\psi_1 + 4(\Delta x)^2 & -4 & \cdots & 0 & 0 & 0 & \cdots & 0 & 0 & 0 \\
\vdots & \vdots & \ddots & \vdots & \vdots & \vdots & \ddots & \vdots & \vdots & \vdots \\
0 & 0 & \cdots & H_{\mathrm{r},1} & H_{\mathrm{r},2} & H_{\mathrm{r},3} & \cdots & 0 & 0 & 0 \\
\vdots & \vdots & \ddots & \vdots & \vdots & \vdots & \ddots & \vdots & \vdots & \vdots \\
0 & 0 & \cdots & 0 & 0 & 0 & \cdots & -2(\Delta x)^2 & H_{\mathrm{r},4} & H_{\mathrm{r},5} \\
0 & 0 & \cdots & 0 & 0 & 0 & \cdots & 0 & H_{\mathrm{r},6} & H_{\mathrm{r},7}
\end{bmatrix}
$$

$$
H_{\mathrm{r},1} = (\Delta x)^2 \cdot \left(\frac{1}{2m} - 1\right) \cdot \left(\frac{2}{\eta_{\mathrm{m,m-1}} + 1}\right)
$$

$$
H_{\mathrm{r},2} = \psi_{\mathrm{m,n}} + 2(\Delta x)^2 + (\Delta x)^2 \cdot \left(\frac{1}{2m} - 1\right) \cdot \left(\frac{\eta_{\mathrm{m,m-1}} - 1}{\eta_{\mathrm{m,m-1}} + 1}\right) + (\Delta x)^2 \cdot \left(\frac{1}{2m} + 1\right) \cdot \left(\frac{\eta_{\mathrm{m+1,m}} - 1}{\eta_{\mathrm{m+1,m}} + 1}\right)
$$

$$
H_{\mathrm{r},3} = (\Delta x)^2 \cdot \left(\frac{1}{2m} + 1\right) \cdot \left(\frac{-2\eta_{\mathrm{m+1,m}}}{\eta_{\mathrm{m+1,m}} + 1}\right)
$$

$$
H_{\mathrm{r},4} = \psi_{\mathrm{M,n}} + 2(\Delta x)^2 + \varphi \cdot (\Delta x)^2 \cdot \left(\frac{1}{2M} + 1\right), H_{\mathrm{r},5} = -\varphi \cdot (\Delta x)^2 \cdot \left(\frac{1}{2M} + 1\right)
$$

$$
H_{\mathrm{r},6} = -h_{\mathrm{c,Fld}} \cdot Pe_{\mathrm{Fld,in}}\, , H_{\mathrm{r},7} = \psi_{\mathrm{Fld}} + h_{\mathrm{c,Fld}} \cdot Pe_{\mathrm{Fld,in}}
$$

A.9 Park Transformation

Considering phase a alignment to q-axis, the Park transformation (three-phase abc to dq-axis) is shown in equation (9.32) [162].

$$\begin{bmatrix} d \\ q \\ 0 \end{bmatrix} = \frac{2}{3} \begin{bmatrix} \sin(\theta) & \sin(\theta - \frac{2\pi}{3}) & \sin(\theta + \frac{2\pi}{3}) \\ \cos(\theta) & \cos(\theta - \frac{2\pi}{3}) & \cos(\theta + \frac{2\pi}{3}) \\ \frac{1}{2} & \frac{1}{2} & \frac{1}{2} \end{bmatrix} \times \begin{bmatrix} a \\ b \\ c \end{bmatrix} \tag{9.32}$$

To convert the $dq0$ frame values to the three-phase abc system, the inverse Park transformation is used. This transformation is shown in equation (9.33) .

$$\begin{bmatrix} a \\ b \\ c \end{bmatrix} = \begin{bmatrix} \sin(\theta) & \cos(\theta) & 1 \\ \sin(\theta - \frac{2\pi}{3}) & \cos(\theta - \frac{2\pi}{3}) & 1 \\ \sin(\theta + \frac{2\pi}{3}) & \cos(\theta + \frac{2\pi}{3}) & 1 \end{bmatrix} \times \begin{bmatrix} d \\ q \\ 0 \end{bmatrix} \tag{9.33}$$

Karlsruher Schriftenreihe zur Supraleitung

Karlsruher Institut für Technologie (KIT) | ISSN 1869-1765

Band 001 **Christian Schacherer**
Theoretische und experimentelle Untersuchungen zur Entwicklung supraleitender resistiver Strombegrenzer.
ISBN 978-3-86644-412-6

Band 002 **Alexander Winkler**
Transient behaviour of ITER poloidal field coils.
ISBN 978-3-86644-595-6

Band 003 **André Berger**
Entwicklung supraleitender, strombegrenzender Transformatoren.
ISBN 978-3-86644-637-3

Band 004 **Christoph Kaiser**
High quality Nb/Al-AlOx/Nb Josephson junctions. Technological development and macroscopic quantum experiments.
ISBN 978-3-86644-651-9

Band 005 **Gerd Hammer**
Untersuchung der Eigenschaften von planaren Mikrowellenresonatoren für Kinetic-Inductance Detektoren bei 4,2 K.
ISBN 978-3-86644-715-8

Band 006 **Olaf Mäder**
Simulationen und Experimente zum Stabilitätsverhalten von HTSL-Bandleitern.
ISBN 978-3-86644-868-1

Band 007 **Christian Barth**
High Temperature Superconductor Cable Concepts for Fusion Magnets.
ISBN 978-3-7315-0065-0

Die Bände sind unter www.ksp.kit.edu als PDF frei verfügbar oder als Druckausgabe bestellbar.

Band 008 **Axel Stockhausen**
Optimization of Hot-Electron Bolometers for THz Radiation.
ISBN 978-3-7315-0066-7

Band 009 **Petra Thoma**
Ultra-fast $YBa_2Cu_3O_{7-x}$ direct detectors for the THz frequency range.
ISBN 978-3-7315-0070-4

Band 010 **Dagmar Henrich**
Influence of Material and Geometry on the Performance of Superconducting Nanowire Single-Photon Detectors.
ISBN 978-3-7315-0092-6

Band 011 **Alexander Scheuring**
Ultrabreitbandige Strahlungseinkopplung in THz-Detektoren.
ISBN 978-3-7315-0102-2

Band 012 **Markus Rösch**
Development of lumped element kinetic inductance detectors for mm-wave astronomy at the IRAM 30 m telescope.
ISBN 978-3-7315-0110-7

Band 013 **Johannes Maximilian Meckbach**
Superconducting Multilayer Technology for Josephson Devices.
ISBN 978-3-7315-0122-0

Band 014 **Enrico Rizzo**
Simulations for the optimization of High Temperature Superconductor current leads for nuclear fusion applications.
ISBN 978-3-7315-0132-9

Band 015 **Philipp Krüger**
Optimisation of hysteretic losses in high-temperature superconducting wires.
ISBN 978-3-7315-0185-5

Band 016 **Matthias Hofherr**
Real-time imaging systems for superconducting nanowire single-photon detector arrays.
ISBN 978-3-7315-0229-6

Band 017 **Oliver Näckel**
Development of an Air Coil Superconducting Fault Current Limiter.
ISBN 978-3-7315-0526-6

Band 018 **Christoph M. Bayer**
Characterization of High Temperature Superconductor Cables for Magnet Toroidal Field Coils of the DEMO Fusion Power Plant.
ISBN 978-3-7315-0605-8

Band 019 **Shengnan Zou**
Magnetization of High Temperature Superconducting Trapped-Field Magnets.
ISBN 978-3-7315-0715-4

Band 020 **Ilya Charaev**
Improving the Spectral Bandwidth of Superconducting Nanowire Single-Photon Detectors (SNSPDs).
ISBN 978-3-7315-0745-1

Band 021 **Juliane Raasch**
Electrical-field sensitive $YBa_2Cu_3O_{7-x}$ detectors for real-time monitoring of picosecond THz pulses.
ISBN 978-3-7315-0786-4

Band 022 **Yingzhen Liu**
Design of a superconducting DC wind generator.
ISBN 978-3-7315-0796-3

Band 023 **Sebastian Hellmann**
Research and Technology Development on Superconducting Current Limiting Transformers.
ISBN 978-3-7315-0804-5

Band 039 **Ali Khonya**
System-Level Modeling of the Powertrain for an Electric Aircraft with Superconducting Components.
ISBN 978-3-7315-1435-0